ALSO BY TIMOTHY FERRIS

BOOKS

The Whole Shebang: A State-of-the-Universe(s) Report

The Universe and Eye

The Mind's Sky: Human Intelligence in a Cosmic Context

World Treasury of Physics, Astronomy, and Mathematics
(editor)

Coming of Age in the Milky Way

The Practice of Journalism
(with Bruce Porter)

SpaceShots

Galaxies

Murmurs of Earth: The Voyager Interstellar Record
(with Carl Sagan et al.)

The Red Limit: The Search for the Edge of the Universe

FILMS

The Creation of the Universe

Life Beyond Earth

LIFE BEYOND EARTH

TIMOTHY FERRIS

SIMON & SCHUSTER

NEW YORK   LONDON   TORONTO   SYDNEY   SINGAPORE

SIMON & SCHUSTER
Rockefeller Center
1230 Avenue of the Americas
New York, NY 10020

Designed by Joseph Rutt
Printed in China through Four Colour Imports, Ltd.
1   3   5   7   9   10   8   6   4   2
Library of Congress Cataloging-in-Publication Data
Ferris, Timothy.
Life beyond earth / Timothy Ferris.
p.   cm.
Includes index.
1. Life on other planets.   2. Outer space—Exploration.   I. Title.
QB54.F415   2000
576.8'39—dc21   00-038782
ISBN 0-684-84937-2

For Carl Sagan
(1934–1996)

CONTENTS

PREFACE

12

I. ARE WE ALONE?

17

IN THE BEGINNING

19

EVOLUTION

37

THE HABITABLE ZONE

61

THE ORIGIN OF LIFE

93

THE ICE ZONE

99

TERRAFORMING

119

II. IS ANYBODY LISTENING?

123

VISITORS

125

THE INFOSPHERE

139

MESSAGE IN A BOTTLE

167

DESTINY

187

OTHER VOICES

199

UNITS OF MEASUREMENT

210

ACKNOWLEDGEMENTS

211

PHOTO CREDITS

212

INDEX

217

LIFE BEYOND EARTH

PREFACE

People have been speculating about life beyond Earth since the dawn of history. What's new is that we have now begun to acquire the tools necessary to replace some of these speculations with fact. Humans have collected rocks on the surface of the Moon, sent instrumented probes to the planets, employed large telescopes on the ground and in orbit to study the stars, and explored the Earth to investigate how life here began and evolved.

So far, the results of the search for extraterrestrial life have been negative. The Moon, long portrayed in hoax and hopeful hypotheses as inhabited, is lifeless. Venus turned out to be fearfully hot and dry, a disappointment for those who had envisioned tropical jungles hidden beneath its perpetual clouds. Space probes dispatched to Mars have found no clear evidence of biological activity, although it may take decades of further study to determine whether there is—or once was—life on Mars.

Meanwhile, other avenues of research have opened new vistas onto the subject of extraterrestrial life and the related issue of how life began. Here on Earth, living organisms have been found thriving in hot vents in the dark depths of the sea, in frigid Antarctic waters, in core samples drilled miles

Milky Way mosaic shows our galaxy as viewed from within. The center of the galaxy, obscured behind the clouds of dust and gas that line the disk, lies toward the right edge of the image.

down into the rocks, and floating high in the air. These findings show that life is tougher and more adaptive than had been assumed, and imply that life could exist on planets that had been thought to be too hot or cold. Jupiter's satellite Europa, for example, may have a global ocean and hot vents resembling the fertile fonts found in Earth's seas. Meanwhile, astronomers have adduced evidence of planets orbiting stars near the sun, confirming the long-held hypothesis that our galaxy contains billions of planets, some of which could be expected to have life even if the advent of life is rare. Earth scientists are finding new evidence that life began here quite early on, adding fuel to the argument that biology arises in the normal course of things, at least on planets that resemble Earth. Astrophysical studies of nebulae in the Milky Way galaxy and beyond show that the basic ingredients of life as we know it are not unique to our quarter of the cosmos but are widespread: Water, organic molecules, and sources of energy are found in countless locations, suggesting that the universe may have given birth to life in a literally unimaginable variety of environments.

From such discoveries come new incentives to keep learning about our planet and its place in the wider universe. These efforts are not forbiddingly expensive, as such things go—the United States in peacetime spends far more money each year on military applications in space than on manned and unmanned space exploration. Nonetheless they are criticized by those who have reasoned their way to the conclusion that there is no life, or at least none that is intelligent, out among the stars. Their arguments are not wholly without merit. It is true that we have as yet no solid evidence that extraterrestrial life exists, and that dream-struck enthusiasts might want to keep searching even if we had one day scrutinized a million planets and found that all were sterile. (The universe is so vast that we'd have hardly begun; life might yet be found on the next planet, or the one after that.) But we know so little so far that to stop now would be at best premature. We tried *not* searching, for more than ten thousand years; it didn't work. If science has taught us anything, certainly it is that one should never underestimate the scope of human ignorance and its influence on our commonplace ways of thinking.

It may be that the search for life beyond Earth is more about exploration than science, but that hardly diminishes its stature. Exploration has been central to our success as a species, and to cease exploring is to risk becoming less than human. Certainly the potential rewards of finding life elsewhere would be substantial. Science, which thrives on comparison and seldom does well when studying something of which it has but a single example, would benefit immeasurably from knowledge of just a single species of extraterrestrial life with which to compare the one form of biology that we know about. And the influences that would be brought to bear on philosophy, theology, history, and art from the discovery of even a humble life form, much less an intelligent one, can scarcely be overestimated.

Science sits at the center of our society, which deserves to be told about its findings and its informed speculations. It was to this end that I agreed to write and narrate the two-hour documentary film after which this book is titled and from which its words and many of its images are taken. Such films are not textbooks, designed to prepare their audiences to pass biology or astronomy examinations, but they can sensitize people to science and involve them in the enterprise of scientific research, if only as informed generalists. Their aim is not so much to provide answers as to help improve the quality of the questions we all ask, whenever we wonder about how life and intelligence began, who we are, and where we came from. We are rank beginners in this quest, and as such can reap the benefits of keeping an active imagination; as the Zen master Shunryu Suzuki put it, "In the beginner's mind there are many possibilities, but in the expert's there are few." To reflect on the wild diversity of life on this one planet, and not least on the rich diversity of individuals within our one species, is to consider how much we have not yet imagined, but can perhaps begin to learn.

—Timothy Ferris
Rocky Hill Observatory
1/1/2000

I.

ARE WE ALONE?

IN THE BEGINNING

In the beginning, when all was fire,
there were no stars or planets,
no atoms or molecules,
and no life.

Eons passed, and life appeared,
on at least one small planet,
orbiting an average star
in a spiral galaxy called the Milky Way.

On that planet, one species,
endowed with the capacity to think and speak,
has begun to wonder: Did it happen only here?

What follows is a report, from a few members of that species,
on the search for life beyond Earth.

Star-forming region RCW38 (distance ~5,000 light years), obscured by intervening clouds when observed in visible light, can be seen in this image taken at infrared wavelengths by the European Southern Observatory's Very Large Telescope.

All earthly life is kin.
We're all members of the same ship's company, adrift
on a planet that has nurtured life for almost five billion years.

But while life is old, we humans are young.
Since life on Earth began, the Sun and its planets have completed
about a dozen orbits around the center of the Milky Way galaxy.
Since humans came along,
the solar system has wheeled less far
than the second hand of a clock ticks off in a single second.

So, we're young.
But we're also bright—or so we like to think.
Of all the animals on Earth, we alone
have the capacity to investigate how life began,
to wonder where it might be headed, and to ask
whether there are creatures on other planets
who are asking the same kinds of questions.

Right: The spiral galaxy NGC 4414 (distance ~60 million light years), imaged by the Hubble Space Telescope in 1999. The blue colors are produced by giant young stars, yellow by older stars, and reds by glowing clouds of hydrogen gas. The black lines interlaced in the disk are clouds of interstellar dust and gas.

Left: The solar system takes 226 million years to complete one orbit around the center of the Milky Way galaxy. (The Sun moves at 135 meters per second; the circumference of its galactic orbit is 163,000 light years.) In the ~200,000 years since Homo sapiens appeared on Earth, it has moved less far than a fine line drawn in the sand on this sketch of the galaxy.

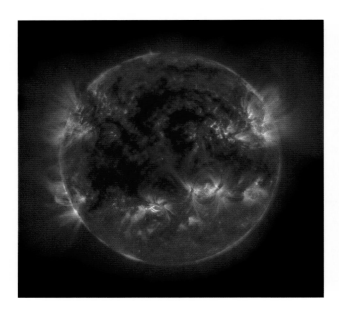 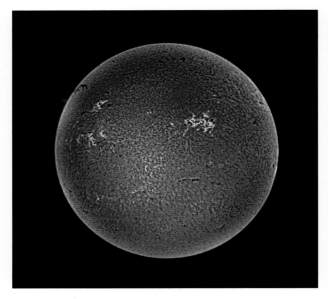

The Sun (diameter 1.4 million km, volume sufficient to hold 1.3 million Earths) imaged (left) in the high-energy wavelengths of extreme ultraviolet light by SOHO, the Solar and Heliospheric Observatory satellite, and (right) in hydrogen alpha light by the amateur astronomer Robert Gendler.

Our quest for life beyond Earth begins
by searching the cosmos for the three things that terrestrial life requires,
energy, water, and organic molecules—
molecules that contain carbon, the basis of life.
Scientists are now learning that all three ingredients
are not unique to Earth,
but are found throughout the universe.

Most of Earth's energy comes from the sun,
but the Sun is just one among a billion trillion stars.
There's plenty of energy out there.

Opposite: The solar corona becomes visible during total eclipses of the sun, which is made possible by the remarkable coincidence that the disks of the Moon and the Sun have almost exactly the same apparent size when viewed from Earth. Photographed by John Gleason in Aruba, Dutch West Indies, on February 26, 1998, through a 106-mm refractor telescope.

Water is essential to life, as people have long appreciated.
The Greek philosopher Thales portrayed water as the basis of all things.
Confucius said, "The man of wisdom delights in water."
Human beings are mostly water.
The cells in our bodies are little bags of water and organic molecules.

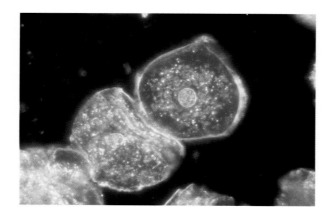

Left: Epithelial cells like these (diameter ~ 40 microns = 400,000 m), from the human mouth, line all the body's hollow structures other than blood and lymph vessels. They have no blood supply of their own, but survive on oxygen and nutrients diffused to them from underlying tissue.

Right: Wahkeenah Falls, Columbia River Gorge, Washington State, 1998. Its rocks, predominantly volcanic in origin, are only ~15 million years old.

Astronomers have found evidence of water and organics
in the solar system, and beyond.
Mercury, the planet closest to the Sun, has ice at its poles,
as does the Moon,
and the planet Mars.

Left: Although Mercury (diameter 4,878 km), the closest planet to the Sun, gets quite hot (470° C), radar studies suggest that there may be water ice in its polar regions. Mosaic of eighteen images taken by the Mariner 10 space probe on March 29, 1974, from a distance of ~200,000 km.

Right: Earth's Moon, photographed by a high-resolution camera mounted aboard the Apollo 15 service module, August 1971. What was a nearly full Moon as seen from Earth looked less than half illuminated from the astronauts' perspective. Recent studies have found evidence of water ice near the lunar poles.

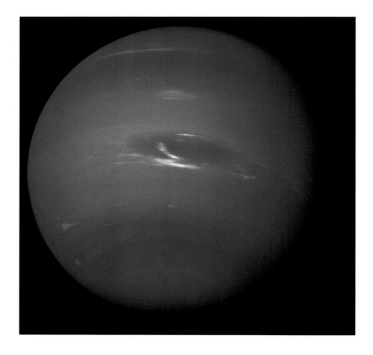

The giant outer planets harbor organics,
and many of their moons are covered with ice.

Swarms of comets orbit the sun.
Each is an iceberg,
freighted with water and organics.

New planets are forming in Orion
from disks of rocks and ice surrounding newborn stars.
These planets incorporate the water and organics in the clouds,
while their central stars bathe them in energy.

Right: Orion Nebula (distance ~1,600 light years; diameter ~30 light years), photographed with the 1.2-m United Kingdom Schmidt telescope at Siding Spring, Australia. Nebulae like these contain enough water and organic materials to support life on hundreds of thousands of Earthlike planets.

Plants growing in a fresh lava field, Kiluea, Hawaii, photographed May 1992.

The energy that sustains life
can come from inside a planet as well.
On geologically active planets like Earth,
volcanoes pump water vapor and carbon compounds
into the atmosphere,
making them available to living organisms.
These veins of glowing lava are the life's blood of a thriving planet.

Old volcanoes—like Yellowstone, the largest volcano in the United States—
are fountainheads of warmth and nourishment
that link the geological activity below to the vitality of life above . . .
on Earth and, perhaps, on other planets as well.

Right: Yellowstone Caldera (overall diameter 80 x 50 km; altitude 2,800 m) last erupted 70,000 years ago. When fully active (0.5–2.0 million years ago), it produced some of the most powerful eruptions in Earth's known geological history.

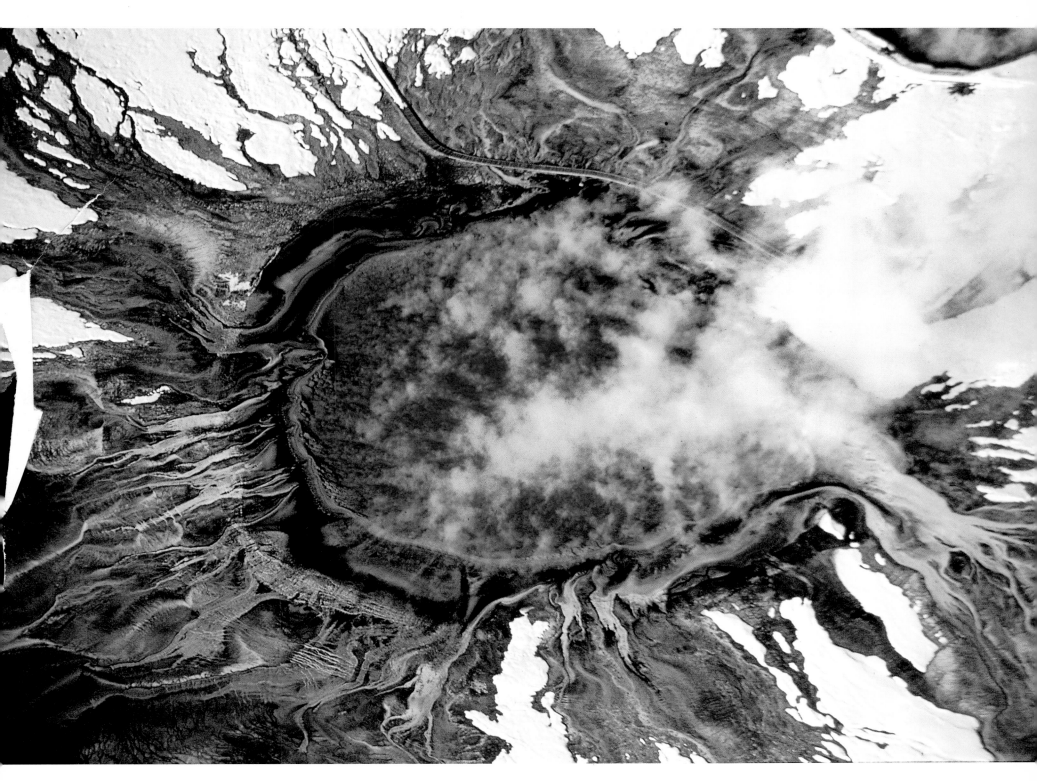

From the center of the Earth to the far-flung galaxies,
we find evidence that life arose from cosmic processes.
The iron in our blood and the calcium in our bones
were made inside stars. All the silver and gold on Earth,
from the vaults of Fort Knox to the wedding rings in the corner pawnshop,
was forged by stars that exploded long ago, seeding the interstellar clouds
from which the Sun and its planets later formed
with tons of diamond and gold.

Terrestrial life is embedded in a cosmic web,
and it seems reasonable to speculate
that life is cosmically commonplace.

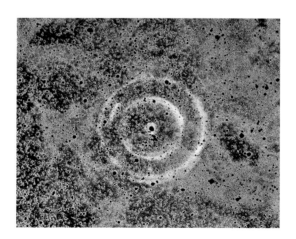

Left: Supernova 1987A, the explosion of an unstable star in a satellite galaxy of the Milky Way called the Large Magellanic Cloud (distance 170,000 light years), was bright enough to be visible to the unaided eye. Light from the explosion, reflected off sheets of interstellar dust 470 and 1,300 light years from the exploding star, formed two rings that appeared after the star itself had faded. Photographed with the 3.9-m Anglo-Australian Telescope in Siding Spring, New South Wales, Australia, February 6, 1989.

Right: Two weeks after its discovery, the supernova was still by far the brightest star in the field of view. A young star, it blew up while still embroiled in the giant nebula from which it had been born. The solar system is thought to have condensed as the result of shock waves generated by a supernova. Photographed by Ray Sharples, March 8, 1987; color negative by David Malin.

EVOLUTION

Once life arose on Earth
it evolved in unpredictable ways.
Biological evolution is not a plodding climb
from "lower" to "higher" forms.
It's a wild, improvisational dance,
as its history reveals.

*Access road, Bonneville
Salt Flats, Utah, looking
toward the future from the
4.5-billion-year marker—
the time when the Earth
cooled sufficiently to form
a solid crust.*

We marked off five kilometers of road to represent the history of the Earth,
on a scale of one kilometer equals one billion years.
The Earth formed about half a kilometer down the road—
some four and a half billion years ago.
The earliest fossil life forms yet identified on Earth
date from 3.75 billion years ago.

These early species are sufficiently sophisticated
to suggest that life had already been evolving for quite a while.
But they don't exactly add up to Saturday night in the big city.
For a very long time, the most complex life forms on Earth
were blobs of single-celled bacteria and algae, called stromatolites.
They may not look like much—indeed, they look a bit like rocks—
but they were the stars of the show for the next three billion years.

That's three kilometers on down the highway.
It would take a while to walk.
We'd better drive.

*Above: Stromatolites,
depicted in a pavement
painting on Bonneville Salt
Flats access road by Rod
Tryon.*

*Right: Porsche C4S
(speed ~280 km/hr)
at Bonneville.*

Even at a hundred miles an hour,
it takes a full minute to traverse the long, dull time
during which life on Earth took on few more complex forms
than bacteria, algae, and plankton.
That's been the dynamic of evolution here on Earth—
not a slow progression from so-called "lower" to "higher" life forms,
but long periods of relative stasis
followed by sudden bursts of innovation.

The most dramatic burst of biological inventiveness came here,
just over half a billion years ago,
when a whole array of creatures equipped
with claws and teeth and tentacles appeared,
in what is aptly called the Cambrian "explosion."
Its cause is something of a mystery,
but the forms taken on by nearly all the organisms on Earth today
represent variations of the plans invented during the Cambrian.
It makes you wonder
just how exotic extraterrestrial life might be.

Left: Cambrian life forms; pavement painting at Bonneville, by Rod Tryon.

Above: Pavement painting of early humans, by Rod Tryon at Bonneville.

Opposite: All recorded history, on the scale of the Highway Through Time (1mm = 5,000 years).

The events that loom largest from our human perspective
all lie along this very last stretch of highway.
Mammals appeared two hundred million years ago—
just two hundred meters from the end of the highway.
Our distant ancestors learned how to walk upright
a little under four million years ago,
and the whole human story lies in the last half meter,
easily overlooked on a nearly five-kilometer
highway through time.

All recorded history—
the rise and fall of empires,
every innovation from the building of the pyramids to the invention of print—
took place in this last few millimeters.
And as for science—well, modern science came along so recently
you'd need a microscope to see it.

Never before has this planet hosted a species acute enough
to find its place on the highway of time,
to ask how life on Earth began,
and to wonder whether life also arose on other planets.

The search for life beyond Earth
is the latest chapter in humanity's long history
of exploring new lands
and finding new varieties of life there.
The islands of the Pacific were settled by navigators
who steered by the stars across dark seas
that must have seemed as vast as deep space.

The circumnavigation of the globe by Magellan's ship *Vittoria*
in the sixteenth century proved that the Earth is a sphere—
unbounded but finite, capable of being mapped and fully explored.

Scientific expeditions like the voyage of Captain James Cook
to the South Pacific, in 1769, had three missions—
to map the earth,
to document the exotic life forms they found,
and to help chart the realm of the other planets.
In Tahiti, Cook and his crew observed a rare transit of Venus across the sun,
from a mountaintop site known ever since as Point Venus.
They were taking part in an effort to measure the size of the solar system.

Undated engraving of Captain James Cook (1728–79) with map and compass, probably after his Venus transit expedition of 1768–71.

*Cook landing at Tanna in
the New Hebrides Islands,
1775. Oil painting by
William Hodges.*

By timing the moment when Venus first impinged on the solar disk
as seen from Tahiti, and from another location thousands of miles away,
the distance from the earth to the sun could be measured, by triangulation.
The relative sizes of all the visible planets' orbits were already known,
so once the size of Earth's orbit was established, that of the other would follow.
The exploration of space had begun,
with Planet Earth its first destination.

Wrote the artist Sydney Parkinson, who accompanied Cook
and painted some of the exotic life forms they encountered in the South Pacific,
"How amazingly diversified are the works of the Deity
within the narrow limits of this globe we inhabit . . ."

Three paintings made by Sydney Parkinson during Cook's Tahiti voyage. Right: Thespesia populnea, *a flower of the seaside Hau tree. Overleaf:* Munida Gregaria, *a small prawnlike lobster;* Macrosystus pyriferus, *a giant kelp.*

Fucus giganteus.

As specimens were brought back to museums
and scientific societies in Paris and London,
it became clear that these newly discovered species of plants and animals were,
for all their diversity, related—
that they shared a common origin.
But if the species are all related, how did they get to be so different?

The brig HMS *Beagle* was a nineteenth-century spaceship,
dispatched on a round-the-world mission to investigate life on Earth.
The naturalist aboard the *Beagle*, Charles Darwin,
was young and inexperienced, fresh out of college.
But he loved collecting plants and animals, and in studying them
he became the first to realize how evolution works—
by combining the creativity of random mutation
with the editing power of natural selection
to turn one species into many.

Crew of the Beagle
*catching a shark off St.
Pauls Rock in the South
Atlantic, 1832.*

HMS Beagle *in the Straits of Magellan, with Mt. Sarmiento in the distance.*

Portrait of Charles Darwin, age forty, by T. H. Maguire, 1849.

Darwin's theory revealed that living things
and the planet they live on
all function under the same set of physical laws.
Ever since Darwin, evolution
has been the pole star of the biological sciences.
"There is grandeur in this view of life," he wrote, "with its several powers,
having been originally breathed by the Creator into a few forms or into one;
and that, whilst this planet has gone cycling on
according to the fixed law of gravity,
from so simple a beginning endless forms most beautiful
and most wonderful have been, and are being, evolved."

Dolphin at Waikaloa,
Hawaii, photographed by
Flip Nicklin, February
1990.

*Sunflower seeds,
photographed in North
Dakota by Layne Kennedy.*

We don't know whether life on other planets
would be based on DNA or some different molecule,
but we do think it would evolve along Darwinian lines.
DNA, the blueprint for life on Earth,
contains a chemical code, based on four polynucleotides,
that harbors all the genetic information necessary
to reproduce the organism to which it belongs.
That's a lot of information.
A single sunflower seed
has the data storage capacity
of a hundred thousand books.

It may seem amazing that all the millions of species of life
could be encoded using just four chemical symbols,
but actually, four is more than enough.
Life could have gotten by with as few as two.
The words and pictures in this book were digitally recorded,
using a code consisting of nothing but zeros and ones.

Regardless of whether extraterrestrial life forms are based on DNA
or some other molecular blueprint,
it is widely presumed that Darwinian selection is the way life works,
not just here on Earth but everywhere else that there is life.

Critics of Darwin's theory say it's impossible to imagine
creating birds or butterflies by randomly organizing four symbols,
just as it's impossible to imagine that randomly scrambling
the letters of the alphabet could produce *King Lear*.

And they're right, so far as that argument goes,
but evolution isn't just random chance.
It's also a matter of discarding variations that do not work
and keeping the ones that do.
The variations that have survived are the ones that have promoted survival.
DNA forms a kind of history library—
a record of perseverance on a changing planet.

Removing letters from existing words produces the word LIFE, illustrating how selection can produce order. High desert near La Cienega, New Mexico, 1998.

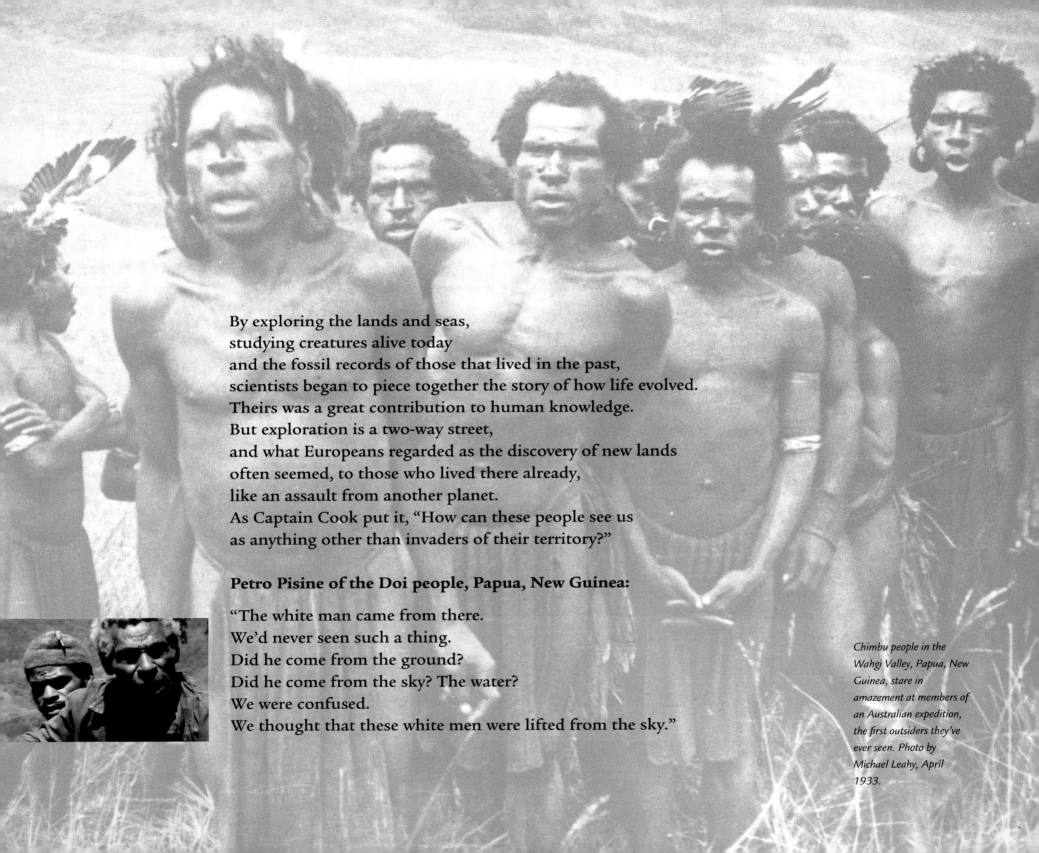

By exploring the lands and seas,
studying creatures alive today
and the fossil records of those that lived in the past,
scientists began to piece together the story of how life evolved.
Theirs was a great contribution to human knowledge.
But exploration is a two-way street,
and what Europeans regarded as the discovery of new lands
often seemed, to those who lived there already,
like an assault from another planet.
As Captain Cook put it, "How can these people see us
as anything other than invaders of their territory?"

Petro Pisine of the Doi people, Papua, New Guinea:

"The white man came from there.
We'd never seen such a thing.
Did he come from the ground?
Did he come from the sky? The water?
We were confused.
We thought that these white men were lifted from the sky."

Chimbu people in the Wahgi Valley, Papua, New Guinea, stare in amazement at members of an Australian expedition, the first outsiders they've ever seen. Photo by Michael Leahy, April 1933.

The English author H. G. Wells noted
that the so-called "discovery" of Tasmania by the Europeans
was "a very frightful disaster" for the Tasmanians.
How would we feel, he wondered,
if technologically advanced creatures from another planet colonized Earth?
Wells's musings about the dark underbelly of colonialism
resulted in the first modern science fiction novel, *War of the Worlds*.

When Orson Welles broadcast a radio dramatization of H. G. Wells's novel,
panic broke out among listeners who took seriously the notion
that the Earth was being invaded by an armada of hostile Martians.
Soon the movie theaters and airwaves were alive with tales of death from above.
In the public imagination, space had ceased to be cold and empty.
It was a jungle out there.

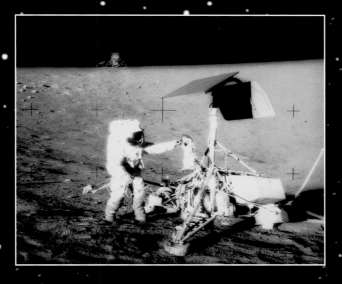 

When humans actually ventured into space,
the search for life was seldom far from their minds.

But where was life to be found?
Our nearest neighbor, the airless Moon, shows no signs of indigenous life,
and rocks returned to Earth by the Apollo astronauts were sterile.

Yet bacteria that were accidentally transported to the Moon
aboard an unmanned lunar lander evidently survived there for years.
And so strong was the association of exploration with the search for life
that when the Apollo astronauts returned to Earth they were placed in quarantine,
just to be sure they hadn't picked up any Moon germs.

THE HABITABLE ZONE

More promising as abodes for life
were the Earth's two neighboring planets, Venus and Mars.
They orbit within what was called the "habitable zone"—
close enough to the sun for solar energy to drive the chemistry of life,
but not so close as to boil off water
or break down the organic molecules on which life depends.
And, it was thought, they might have hot, molten cores
that could power volcanoes like the ones that help sustain life on Earth.

Fresh lava near Kalapana, Hawaii, flowing into the sea (left), and obliterating Harry K. Brown Park (right), 1990. Photos by G. Brad Lewis.

Venus was a favorite of science-fiction writers
who imagined it to be a lush, tropical planet,
a kind of celestial Tahiti.

Perpetually shrouded in clouds, Venus was a mystery.
Nobody knew what was down there.

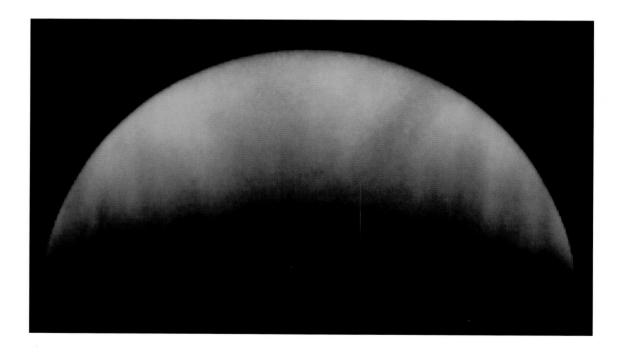

The clouds of Venus,
imaged by the Hubble
Space Telescope in
ultraviolet light, March 21,
1995 (left), and by the
Pioneer Venus Orbiter,
February 5, 1979 (right).

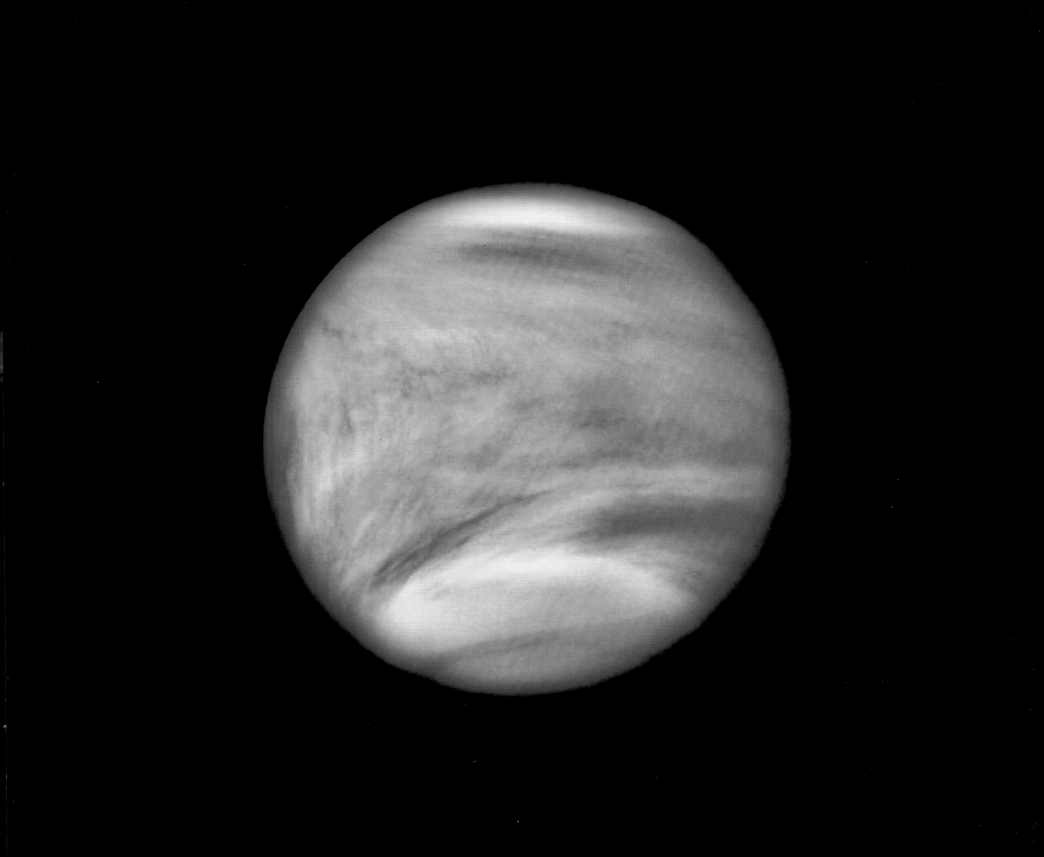

The real Venus proved to be stranger than fiction.
A series of unmanned Venera probes,
dispatched by the Russians from 1966 to 1982,
descended through the dense clouds
and landed on the surface of Venus.
They found, not a lush jungle, but a dry, hellish world,
almost devoid of oxygen, and hot enough to melt lead.
Although the Venera probes were built sturdy as diving bells,
each could take photos and obtain data for only about an hour
before being destroyed by the heat and pressure.
Scientists theorize that Venus fell victim to a runaway "greenhouse effect"—
excess carbon dioxide trapping solar heat under the blanket of its atmosphere.

Many questions remain to be answered about Venus.
Was its climate once more temperate?
And, if so, what went wrong?
Could there have been life there in the past?

*Venera 13 Lander image
of the surface of Venus
east of Phoebe Regio taken
March 1, 1982, during
the 2 hours and 7 minutes
that the probe survived.
Part of the lander appears
at bottom; a jettisoned
camera lens cover lies on
the ground to left of
center.*

David Grinspoon is a rock musician and an astronomer
who works in "comparative planetology"—
using what has been learned about other planets
to enhance our understanding of Earth.

David Grinspoon:

"The surface of Venus remained frustratingly hidden from us for hundreds of years.
In the early 1990s, the Magellan spacecraft,
an American spacecraft equipped with imaging radar,
orbited and imaged the entire planet in stunning detail. All of a sudden
we went from almost no images of the surface of Venus,
except for a few of those tantalizing Russian Lander images,
to a global view in very high resolution of the entire planet.
Early telescopic observers of Venus correctly deduced
that the fuzzy appearance of Venus in telescopic images
is due to the fact that it's completely shrouded in clouds.
Pretty recently we started to get modern scientific data about the planet
which revealed the clouds to be made not of water
but of concentrated sulfuric acid, better known as battery acid.

Above: Astronomer David Grinspoon, playing a "Zen Drum" synthesizer, Denver, Colorado, October 1999. Photo by Tony Read.

Right: Radar image of Venus, compiled from Magellan space probe data radioed to Earth 1990–94. Colors represent elevations. Maximum resolution ~3 km.

"Venus is alive with volcanism.
Other than the Earth, it's the most volcanically active planet in the inner solar system,
and the range of styles of volcanism we see there is stunning.
These pancake domes are made out of a very sticky kind of lava.
That tells us that these are not made out of basalt. Basalt is runny.
This stuff is some kind of lava that's probably more rich in silica,
and less rich in iron and magnesium,
and, therefore, doesn't flow as easily.

"One of the most surprising and controversial findings that we got from Magellan
is that something radical changed relatively recently.
About five hundred million years ago, Venus essentially turned itself inside out,
with volcanic floods that covered up most of the previous surface features,
and probably flooded the air with greenhouse gases,
causing rapid change to the climate as well as to the surface.
This has disturbing implications. It tells us that Earthlike planets
can undergo dramatic changes, in both their surfaces and their climates,
in a relatively recent geologic epoch. And this information could mean
that Earth may have future threatening changes of this magnitude in store for us."

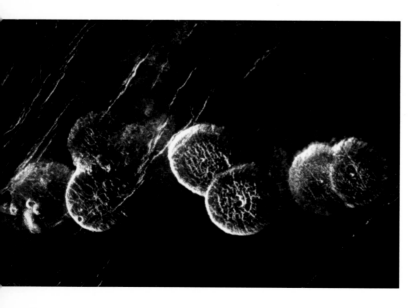

Overleaf: Magellan radar computer—generated perspective view of western Eistla Regio, Venus, September 27, 1991. The large volcano on the left is Gula Mons (height 3 km); on the right, Sif Mons (diameter 300 km).

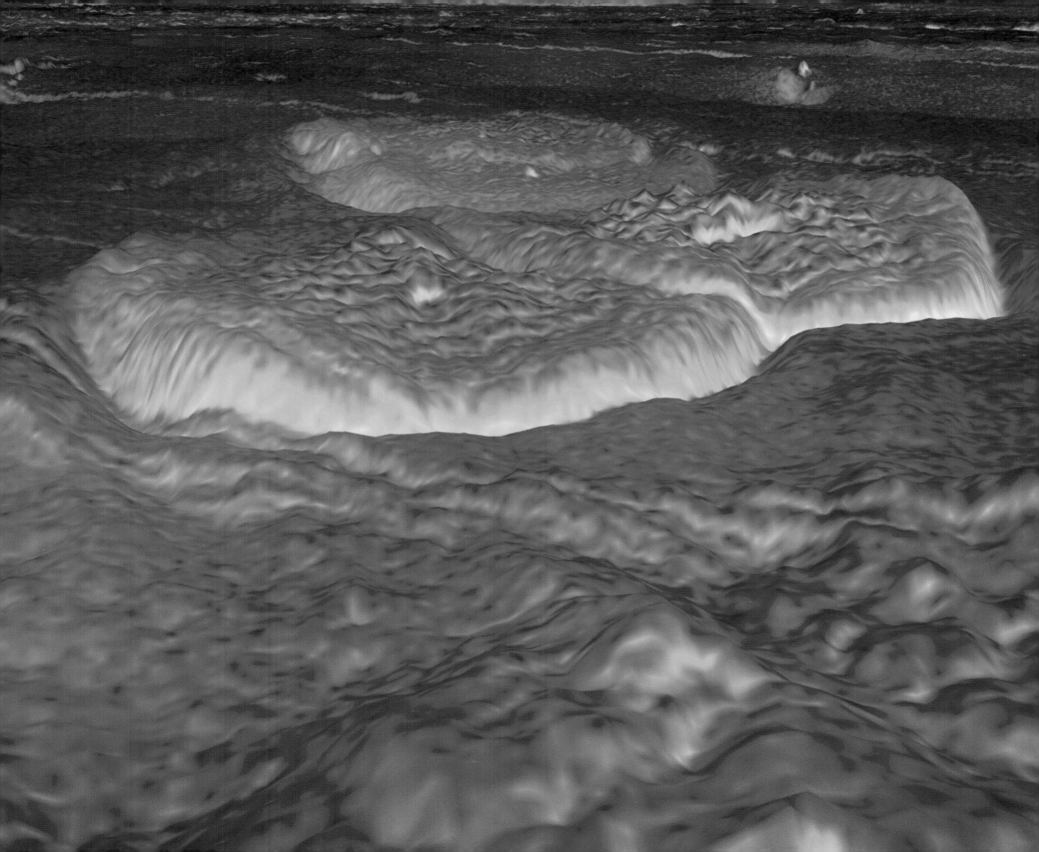

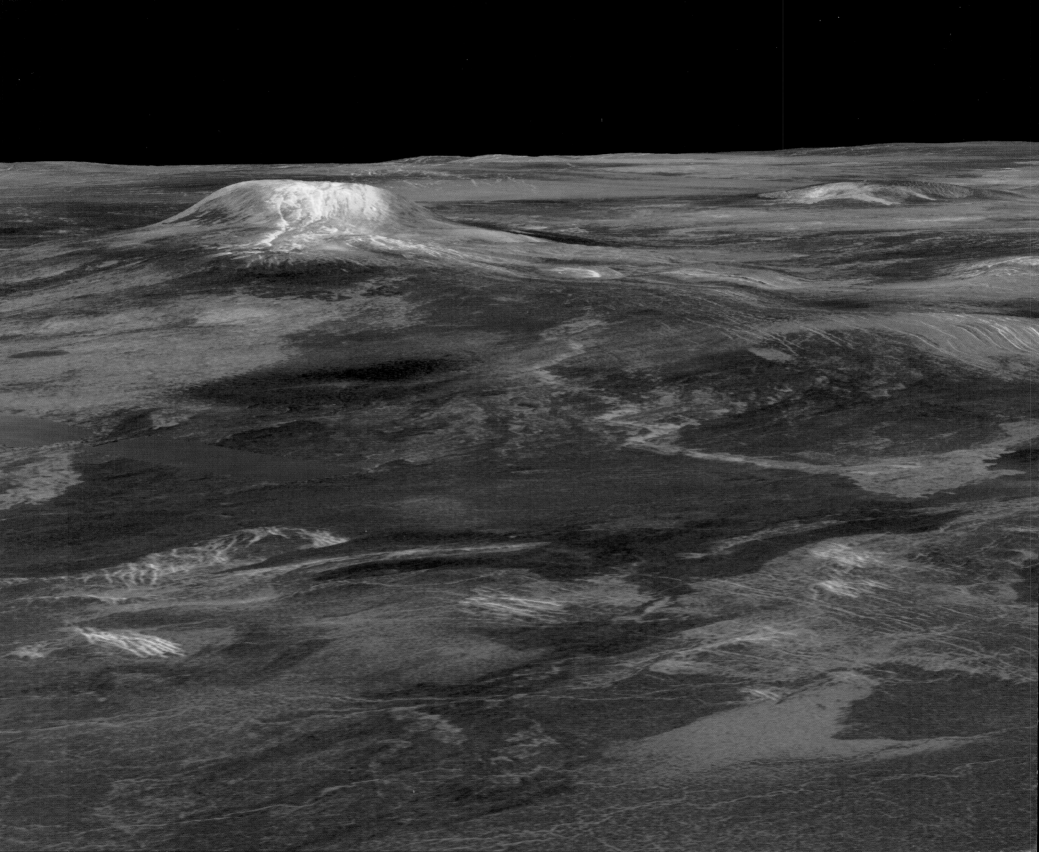

Mars was the planet thought most likely to harbor life.
It lies within the habitable zone,
and its polar caps suggested, even to early observers,
that there could be water there.

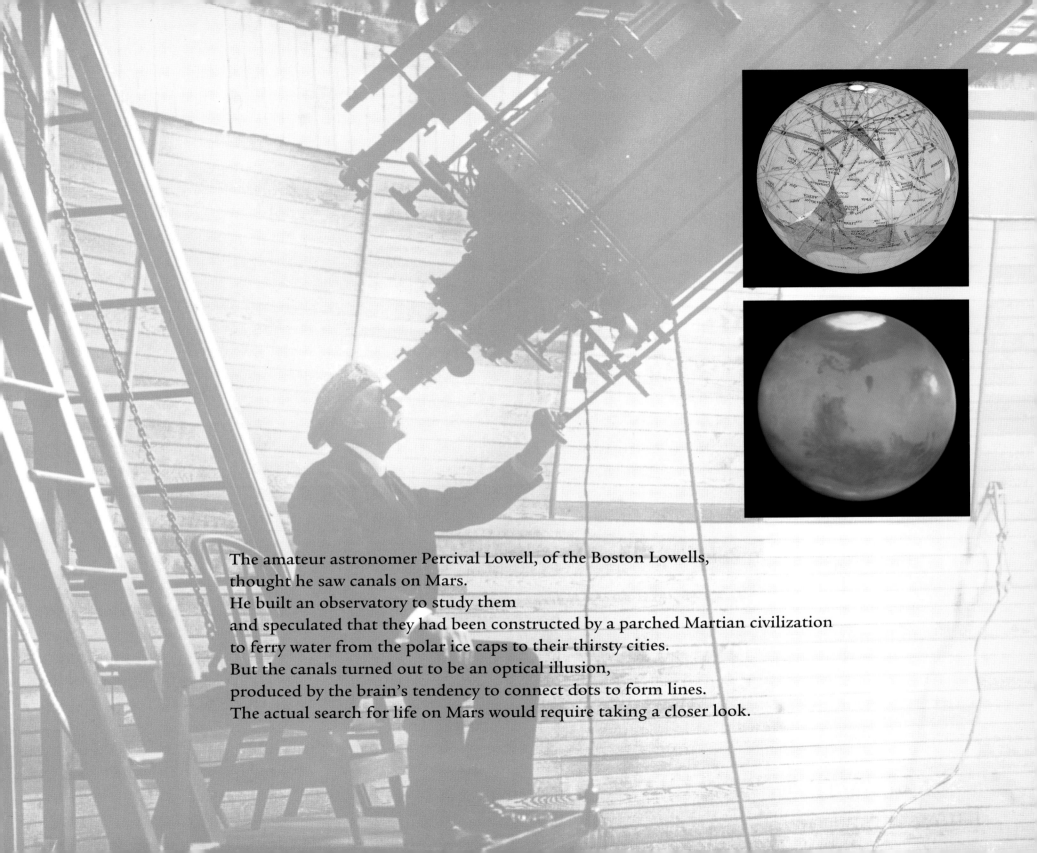

The amateur astronomer Percival Lowell, of the Boston Lowells,
thought he saw canals on Mars.
He built an observatory to study them
and speculated that they had been constructed by a parched Martian civilization
to ferry water from the polar ice caps to their thirsty cities.
But the canals turned out to be an optical illusion,
produced by the brain's tendency to connect dots to form lines.
The actual search for life on Mars would require taking a closer look.

A pair of instrumented Viking space probes were sent to Mars,
looking for signs of life.
Each dispatched a lander to the surface.
In the predawn hours of July 20, 1976,
at NASA's Jet Propulsion Laboratory in Southern California,
scientists and spectators watched anxiously
for humankind's first closeup look at the surface of Mars.

What the pictures revealed was neither any obvious sign of life,
nor a landscape as desolate as that of Venus,
but a different sort of world,
with a character and history all its own.
The landers dug up Martian soil and tested it for signs of biological activity.
The tests found some unexpected chemical reactions,
but no clear evidence of life.

*Right: Portions of Valles
Marineris, imaged by the
Mars Global Surveyor. The
image covers an area
~ 17.3 km wide.*

*Left: Chryse Planitia,
Mars, imaged by the
Viking 1 Lander August
30, 1976. The white
dome in the foreground
covers the lander's
radioisotope power
generator; its high-gain
antenna is at upper right.*

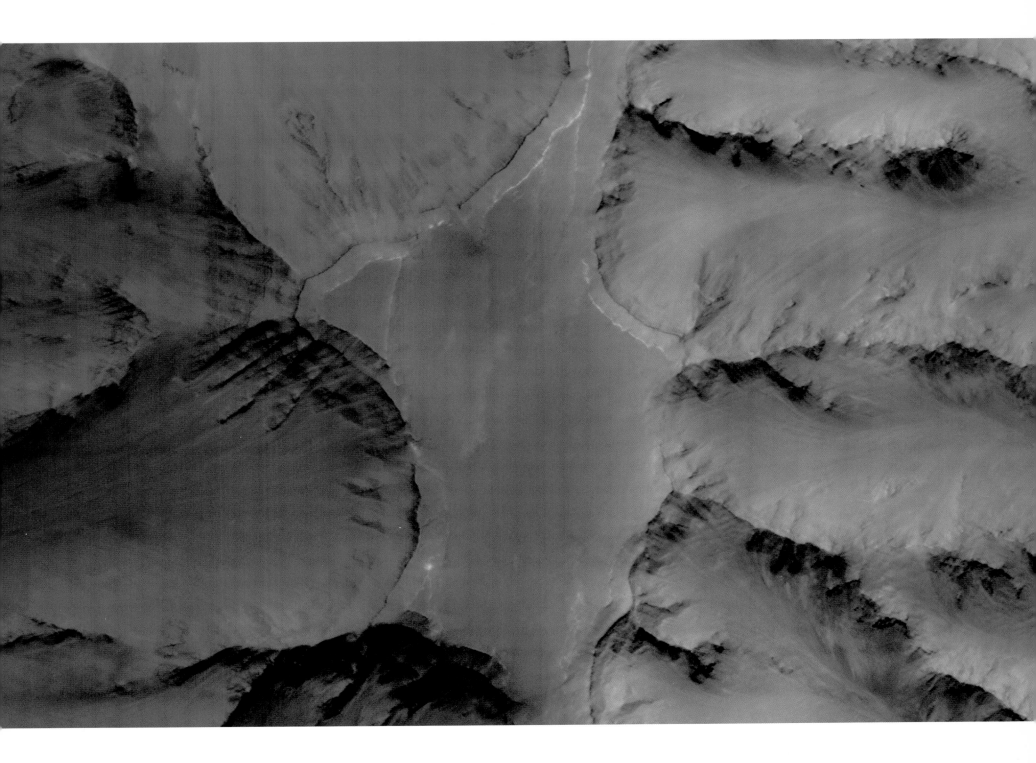

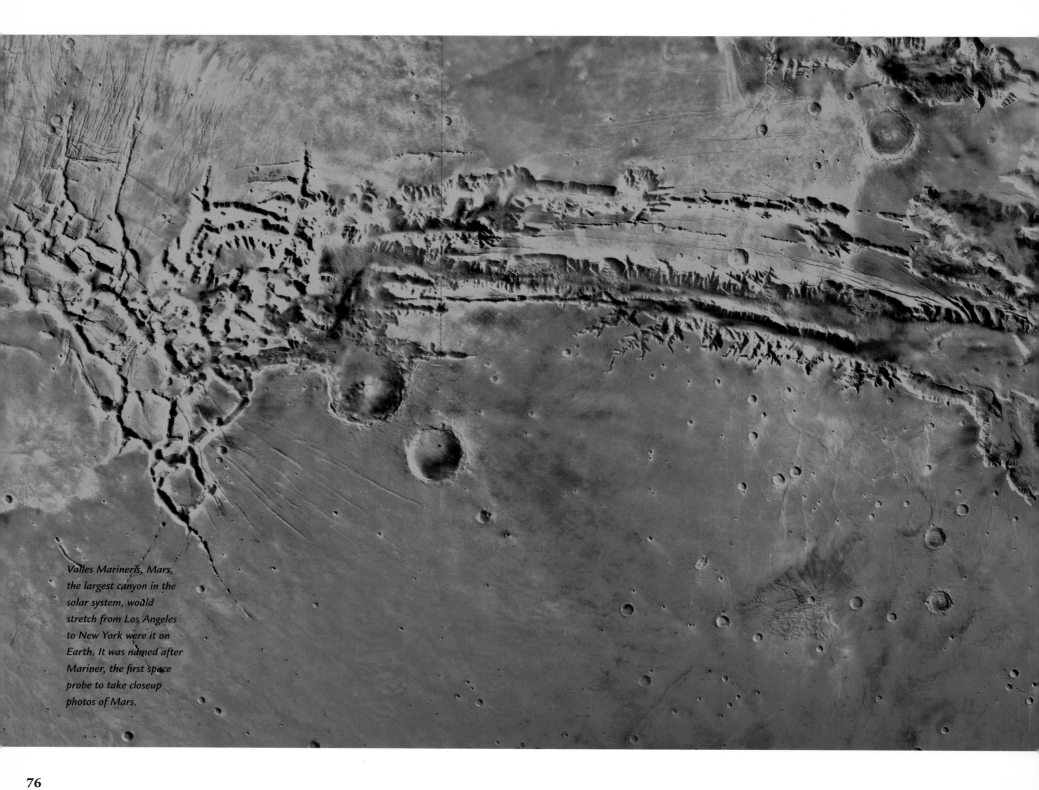

Valles Marineris, Mars, the largest canyon in the solar system, would stretch from Los Angeles to New York were it on Earth. It was named after Mariner, the first space probe to take closeup photos of Mars.

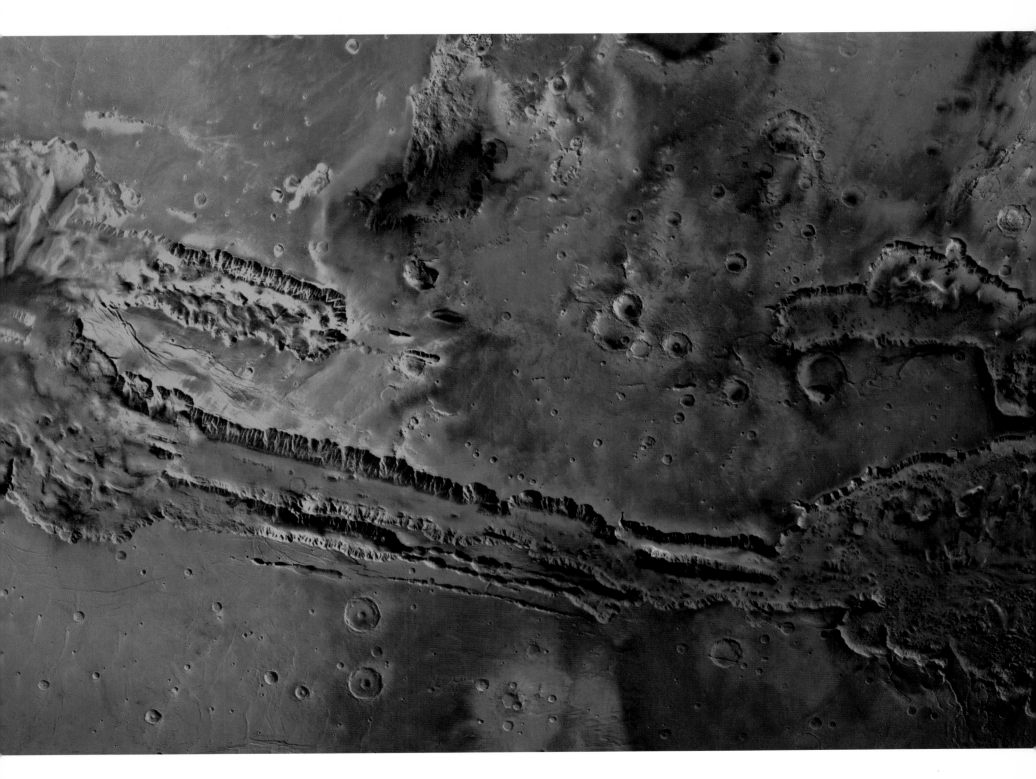

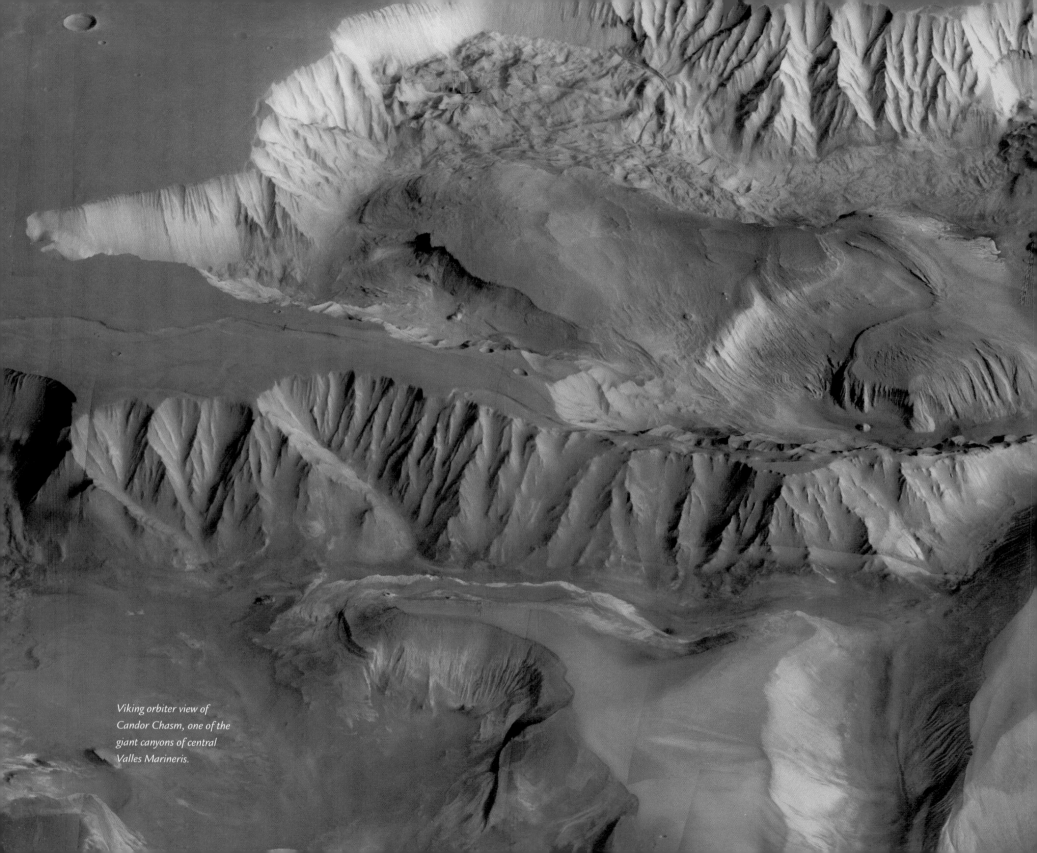

Viking orbiter view of
Candor Chasm, one of the
giant canyons of central
Valles Marineris.

Meanwhile, a pair of Viking orbiters took images that showed dry riverbeds,
indicating that Mars once had liquid water on its surface,
and, therefore, a denser atmosphere—
perhaps one sufficient to support life.

Viking confirmed that Mars, like Venus and Earth, has volcanoes.
This one, Olympus Mons, is the largest volcano in the solar system.
But the very fact that Olympus Mons is so big
suggests that Mars is geologically dead.

Earth's molten core drives plate tectonics.
The plates that support continents and the ocean floor slowly drift,
gigantic barges afloat on liquid stone.
The volcanoes of Hawaii were built, one after another,
from a single hot spot below.
The motion of the surface plates carried each volcano away,
rendering it extinct and leaving the hot spot to build another mountain
in its wake. Mauna Kea, two-thirds the height of Olympus Mons,
is the most recently completed of these colossal construction projects.
But Mauna Kea has since moved off the hot spot,
and it will grow no more.

Meanwhile, the hot spot is building a new volcano to the south,
adding fresh real estate to the island of Hawaii.
And yet another is forming on the ocean floor
—destined, perhaps, to become a Hawaiian island of the future.

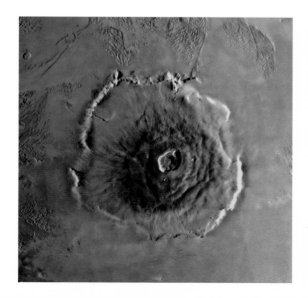
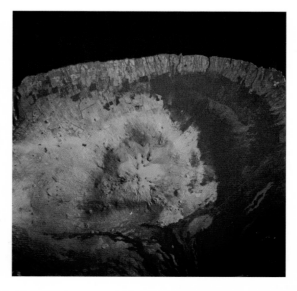

Left: Olympus Mons volcano, Mars (diameter ~600 km, height ~27 km), imaged by the Viking 1 Orbiter, June 22, 1978.

Right: Mauna Kea (altitude from sea level 4,205 m; from ocean floor, 9,750 m), a dormant volcano on the island of Hawaii, photographed from the Space Shuttle in September 1993.

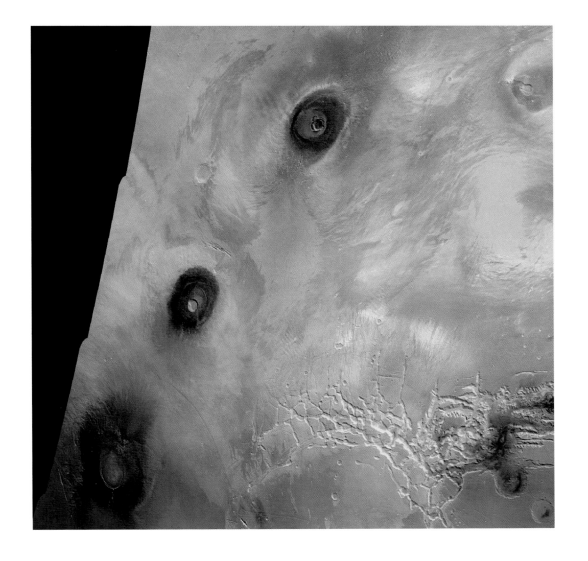

Left: Three Martian volcanoes (from top to bottom, Ascraeus Mons, Pavonis Mons, and Arsia Mons; each ~ 25 km high) in a mosaic of images taken by the Viking 1 Orbiter, February 22, 1980.

Overleaf: Mosaic of Mars, created from 24 images made with the Mars Orbiter Camera on a northern summer day in April 1999. South is at the bottom. Cylindrical projection exaggerates the size of the polar ice cap.

But on Mars, Olympus Mons stayed put,
and grew to great height, indicating
that the core of Mars wasn't hot enough
to move the surface plates around.
The extinct volcanoes of Mars
are geological gravestones.

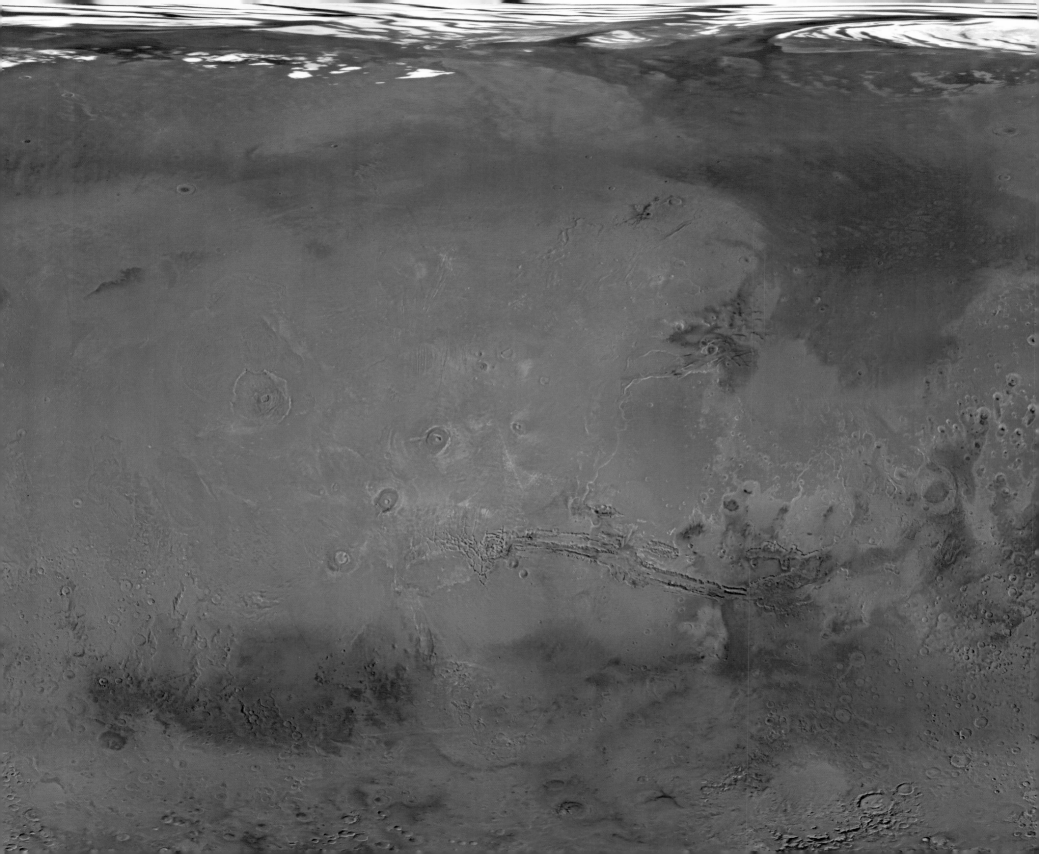

Twenty years after Viking, another probe landed on Mars.
Pathfinder touched down in a dry river delta
and examined rocks carried there from the highlands
in the days when Mars had rivers and streams.
Again, no signs of life were found.

View of the surface of
Mars from the Mars
Pathfinder lander, portions
of which can be seen in the
foreground. The imaging
cameras stand 1.8 m
above the surface.

Long ago, something went terribly wrong on Mars.
The atmosphere thinned out and the planet's ozone layer—if it ever had one—collapsed,
exposing the surface to sterilizing solar ultraviolet light.
The surface waters evaporated, or froze into the ground.

The possibility remains that life got started before this disaster occurred.
If so, did it die out, and leave fossils . . .
or might life still be there, patiently hibernating
in the rusty sands of Mars?

Mars Pathfinder image of the "Twin Peaks," a pair of hills located ~ 1–2 km from the lander. Pathfinder touched down on Mars July 4, 1997.

That life is not conspicuous
doesn't mean it's not there.
This desolate landscape isn't on Mars;
It's a dry valley in Antarctica.

*Upper Wright Valley,
Victoria Land, Antarctica,
photographed by Stuart
Klipper on a 7 x 24-inch
(18 x 60 cm) negative,
February 1992.*

*Living organisms survive
for centuries in these harsh
conditions, the closest on
Earth to the environment
of Mars.*

When the explorer Robert Scott discovered one such valley in 1903,
he wrote that "we have seen no sign of life, not even a moss or lichen."
Yet there is life here.
Slowly growing microorganisms huddle in the frosty soil.

Also to be found in the snows of Antarctica
are pieces of Mars—rocks, knocked off the red planet, that later fell to Earth.
About a dozen such Mars meteorites have been recovered.
In one of them, researchers found what they considered to be fossils of Martian life.
The evidence is preliminary at best,
but it renewed interest in the tantalizing possibility
that living organisms could have traveled around the early solar system,
ferried by rocks that were blown off one planet and wound up on another.

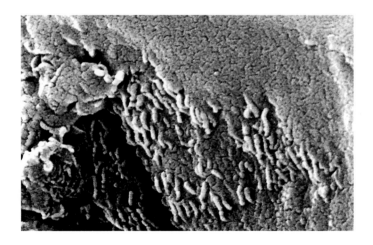

Deep inside a meteorite from Mars (left) tubular structures have been found that some scientists interpret as fossils of primitive life. But these structures are quite small—half the size of the smallest known terrestrial bacteria, and smaller than viruses like the T-lymphocyte (the blue clump in the photo on the right, diameter ~3 microns, magnification 26,400 x). It remains to be seen whether organisms that small actually exist.

THE ORIGIN OF LIFE

One way to investigate the origin of life
is to look for places on the Earth today
where conditions approximate those that pertained back when life began.
But *where* did life begin?
On the ocean floor?
On the surface, in volcanic hot springs?
In tidal pools,
with the first stirrings of evolution prompted, perhaps,
by the rising and falling of tides authored by the then-nearby Moon?
Or in volcanic hot springs, whether up top or on the ocean floor?
And *how* did life begin?

Minerva Terrace hot springs, Yellowstone National Park, October 1987.

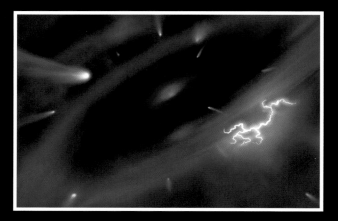

To find out, scientists are investigating the physics
and chemistry of the early solar system.

The Sun's planets formed like slag heaps,
when boulders and gravel orbiting the Sun
were gathered together by gravitational attraction.
Terrestrial life got started so early
that the Earth was still being bombarded by cosmic debris
when life began.

The infant Earth was a violent place,
its atmosphere a strange brew of methane and ammonia.
Life has since refashioned the planet, making it more hospitable.
And that's one of the problems about investigating how life originated.
One has to think of the Earth, not as it is now, but as it was then.
Information stored in the genes of living creatures today
suggests that life evolved from a single heat-loving ancestor—
an organism that could cope with the volcanoes,
sulfuric geysers, and meteorite storms of the early days.
To learn what that ancestor was
could bring us closer to understanding the origin of life.

Large meteorites impact on the young Earth, and the Moon—which formed nearby, and has since receded—looms large in the sky. Traces of the protoplanetary disk can still be seen near the Sun. Don Davis illustration.

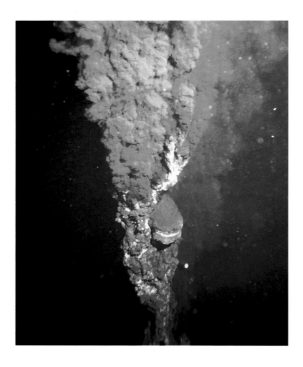

The heat-loving bacteria found in hot springs
are among the oldest organisms on Earth—
a clue that the earliest terrestrial life may have been nourished by the heat,
not of the Sun, but of thermal vents generated by the Earth's molten core.
"Black smokers" are hot volcanic vents on the ocean floor.
They host a rich variety of living organisms that get along in the darkness,
sustained by geothermal energy.

Some creatures like it cold.
On the bottom of the Gulf of Mexico, researchers were startled to find
living worms embedded in methane ice.

97

THE ICE ZONE

The discovery that life can thrive in icy cold and hellish heat
means that life need not have begun in a sunny, placid pond,
as some had imagined, but might have arisen
in the eternal night of the ocean depths—
on Earth, or any other world that has oceans and a molten core,
no matter how far it is from the sun.
So the habitable zone may be larger than anyone had realized.
If the hot cores of geologically active worlds
can sustain life in darkness,
life could exist even among the icy outer planets,
where sunlight is weak.

Jupiter, five times farther from the Sun than the Earth is,
receives only one-twenty-fifth as much sunlight.
So Jupiter's satellites are cold—at least, on the outside.

Jupiter (equatorial diameter 71,398 km) and two of its satellites— Europa (above) and Ganymede (below), imaged January 17, 1979, by the Voyager 1 space probe from a distance of 47 million km. This was one of the first images from which the existence of ice on Europa was inferred.

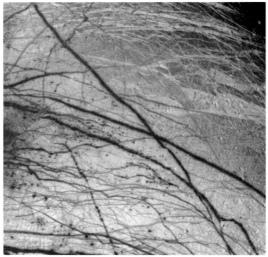

Yet the icy surface of Jupiter's satellite Europa
is marked by lanky fissures—
cracks that suggest the presence of an ocean below
that warms and replenishes the ice.
Europa may have a molten core
that keeps the ocean from freezing
and that could power life-sustaining deep-sea vents
as do Earth's black smokers.

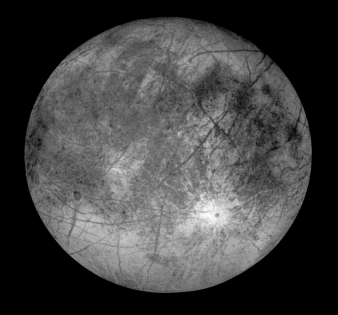

Europa (diameter 3,160 km, about that of Earth's Moon) in true color. The cracks in the ice crust measure up to 3,000 km in length, with a few stretching even farther. The bright impact crater toward the lower right (diameter ~ 50 km), named Pwyll, for the Celtic god of the underworld, formed in recent geological time. Galileo space probe image, September 7, 1996, range 677,000 km.

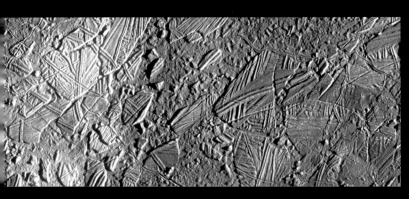

Above: A small region
(~70 x 30 km) of the
thin, disrupted, ice crust
of Europa, with colors
enhanced to bring out
detail. The darker areas
have been blanketed by
dust ejected when a
meteor impact created
the crater Pwyll, 1,000
km to the south. The
small craters in this area
(diameters ~ 500 m)
probably came from
blocks of ice that fell from
the same impact. Galileo
space probe image,
February 20, 1997.

Below: Enhanced-color
photomosaic of the
northern hemisphere of
Europa, imaged in 1998
by the Galileo space
probe, covering an area
800 km x 350 km. The
blue background is water
ice, and cracks indicating
geological activity in the
ice cross the surface. The
brown spots and ridges
may contain mineral
salts. The smallest visible
objects are ~ 230 m in
diameter.

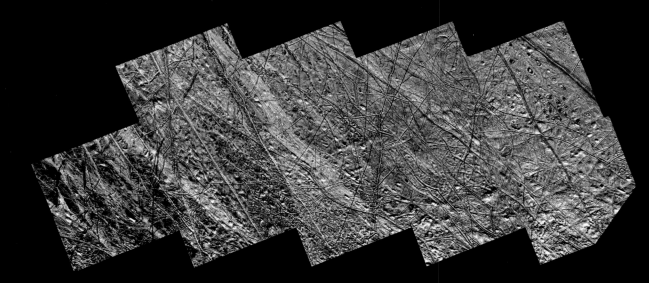

We could find out,
by dispatching a lander to melt a hole in the ice
and drop a submarine into the Europan sea.
The automated submarine could roam Europan's ocean for years,
searching for signs of life.

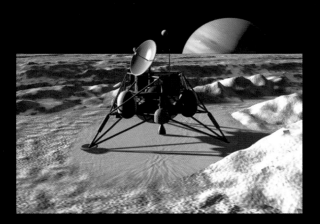

Top: Artist's conception of a robotic submarine exploring a thermal vent on the floor of Europa's putative global sea.

Bottom: A proposed Europa lander begins melting a hole in the ice sheath (estimated thickness >1km) to release a submarine probe into the ocean beneath. Don Davis illustration.

Farther out in the solar system
are other icy moons that are cold on the outside,
but may be hot inside—
among them Neptune's satellite Triton
and Saturn's satellite Enceladus.

By sending probes to these mysterious moons,
scientists can both search for life
and study the primitive environments in which life arises.

Left: Enceladus (diameter 510 km), a satellite of planet Saturn, imaged by the Voyager 2 space probe at a distance of 112,000 km. The surface of Enceladus appears to be covered by young ice.

Opposite: Triton, the largest of Planet Neptune's satellites (diameter 3,800 km) and one of the coldest spots in the solar system, is a site of suspected nitrogen gas volcanoes and methane and nitrogen ice. Mosaic of images obtained by Voyager 2, August 25, 1989.

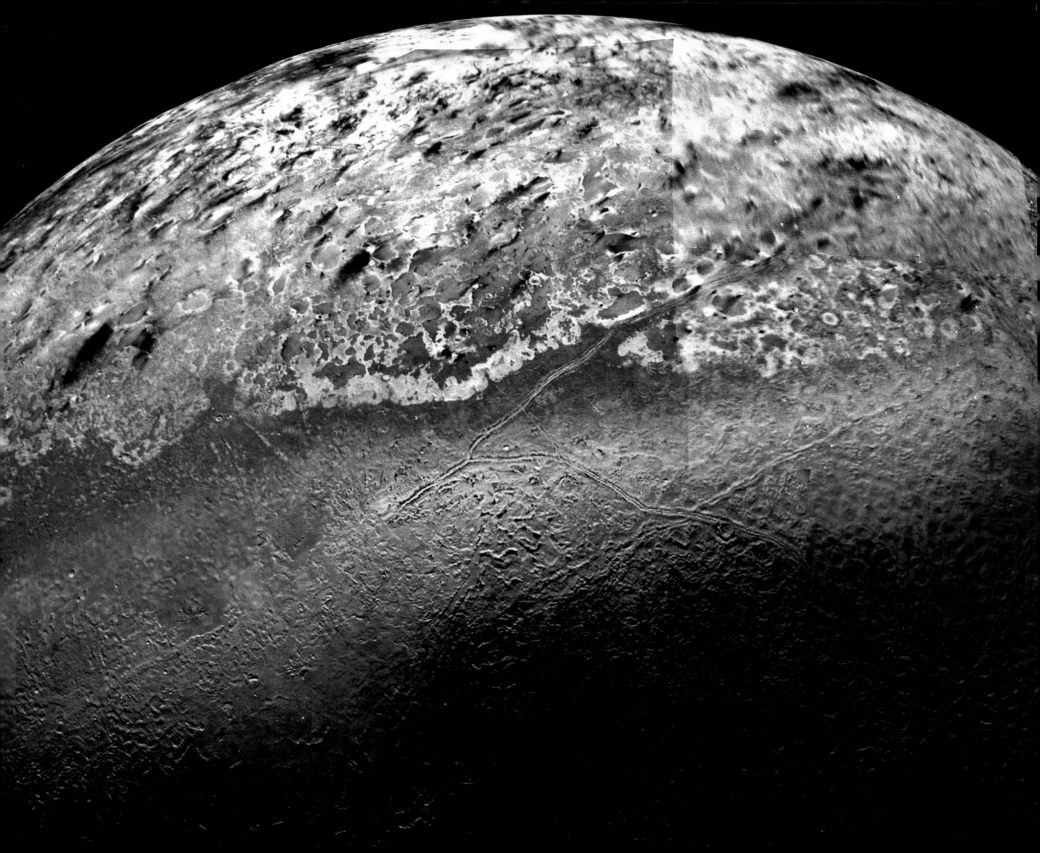

On Titan, the largest satellite of Saturn,
a gentle rain of organic molecules is thought to fall,
amid conditions resembling those of the early Earth.
A probe dispatched by the unmanned Cassini mission to Saturn
was designed to search for clues to life's origins
hidden beneath Titan's clouds.

Far Left: Artist's conception of a view of Saturn through a rare hole in the cloudy atmosphere of its largest satellite, Titan. Don Davis illustration.

Left: Night side of Titan (diameter 5,150 km), showing sunlight diffused through its dense atmosphere. Imaged by Voyager 2 at a range of 907,000 km, August 25, 1981.

Above: The rings of planet Saturn, composed of snowballs and ice, are so thin (~20 m) that they almost disappear from terrestrial view when seen edge-on. In this view, taken with the Hubble Space Telescope in 1995, Titan casts a shadow on Saturn's gaseous disk.

Left: Io (diameter 3,630 km) imaged by the Galileo space probe from a distance of 487,000 km on September 7, 1996.

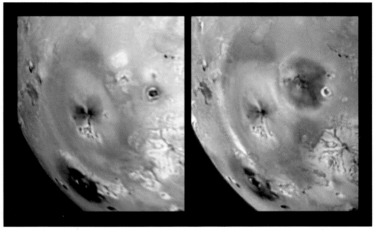

Above: Volcanic eruptions cause rapid changes on the surface of Io, here captured in two images taken by the imaging system on the Galileo space probe on April 4 (left) and September 19, 1997 (right). The new dark spot (diameter 400 km) consists of material ejected from Io's volcano Pillan Patera.

Jupiter's moon Io is a palpitating bag of lava.
Scores of volcanoes pierce its grapeskin-thin surface.
The Voyager spacecraft imaged a volcanic plume shooting into space from Io,
and so, decades later, did the Galileo probe.

These two images of the same spot on Io
show a Texas-sized black splotch of volcanic eruption on the right
that wasn't there when the photo on the left was taken,
just five months earlier.
Observations like these reveal that the geology of a volatile world like Io
can be almost as changeable as the weather on Earth.

The potential expansion of the habitable zone here in the solar system
implies that life might thrive in other places
that had been thought to be uninhabitable.

There could be living creatures afloat in the atmospheres
of gaseous planets like Jupiter . . .

The upper atmosphere of Jupiter, including the Great Red Spot, a cyclone more than 300 years old, imaged by Voyager 1 February 25, 1979, at a range of 9.2 million km. Highest resolution is 160 km. Scientists have speculated that life could exist on Jupiter, perhaps in the form of floating jellyfish that ride in the clouds.

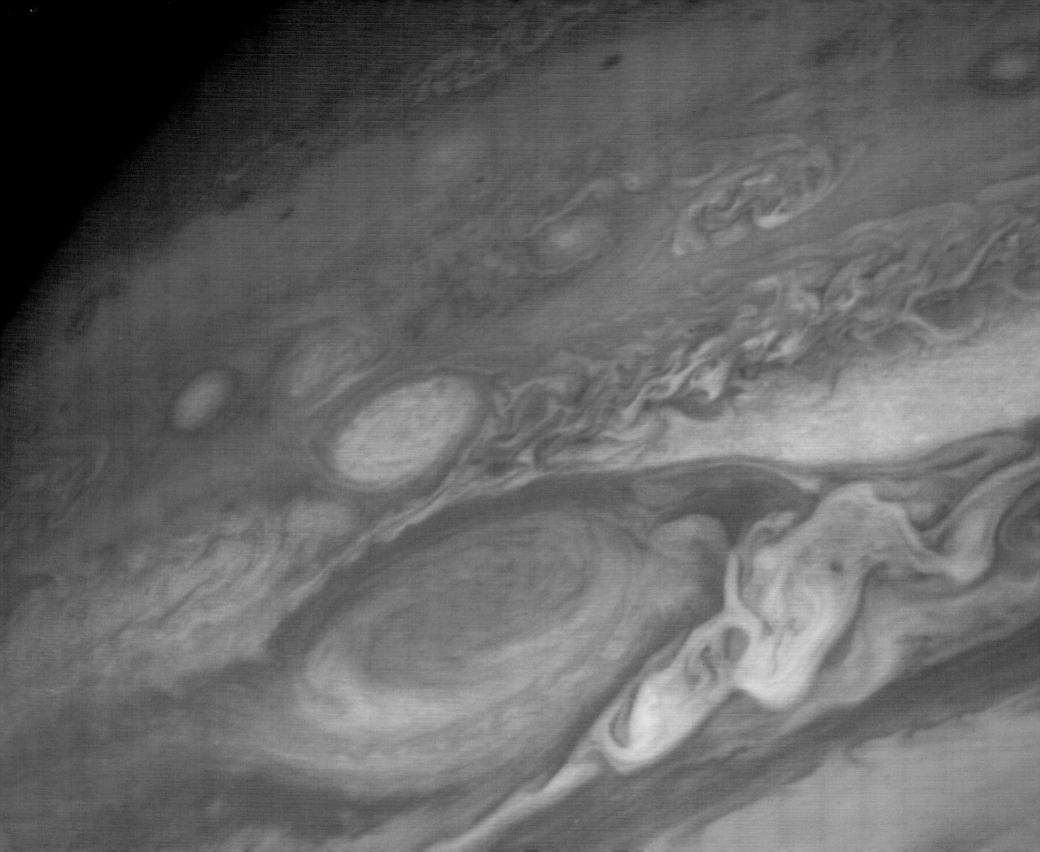

There could be life on young planets still entangled in the dark clouds from which they formed . . . or, conceivably, adrift in the clouds themselves.

... On the ancient planets of old, steadily burning stars in globular clusters ...

The globular star cluster
Omega Centauri (distance
~17,000 light years)
contains several million
stars, some of them among
the oldest in our galaxy.
Photographed by Steve Lee
with the Anglo-Australian
telescope, July 30, 1992.

. . . Or amid the fiery star-forming regions of colliding galaxies . . .

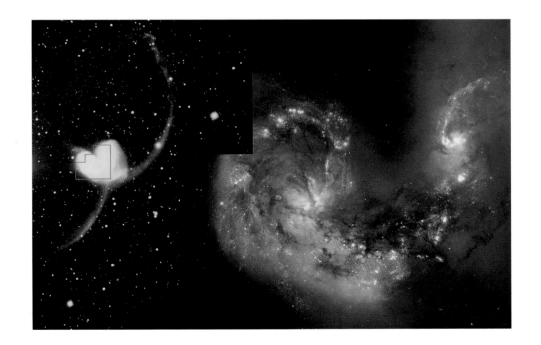

A fireworks show of star formation, touched off when two galaxies collided, has produced over a thousand new star clusters in the galaxies (NGC 4038/4039; distance ~90 million light years; diameters ~100,000 light years) known as the "Antennae" because of their long, uncoiling tails, produced by their gravitational interaction and only partly visible within the frame of this photograph. Such episodes produce many giant stars, which shine brilliantly but die out when only about 1 percent the age of mainstream stars like the Sun. Hubble Space Telescope image.

There could be more life out there than we've ever imagined—
for if the universe has taught us anything, it is that reality
is richer and more resourceful than our wildest dreams.

Left: Gravitational lensing, the warping of space by massive objects that bends light rays coming to us from behind them, wreathes this cluster of galaxies (Abell 2218, distance ~ 3 billion light years) in thin arcs made of starlight. Each is the distorted and magnified image of a galaxy even farther away, the study of which aids in learning how the universe has evolved. Hubble Space Telescope image.

Right: The two giant elliptical galaxies near the center of this image form the core of a cluster of galaxies (distance ~ 250 million light years) in the southern-hemisphere constellation Norma. A more distant and as yet unmapped cluster is pulling our galaxy in this direction. European Southern Observatory image.

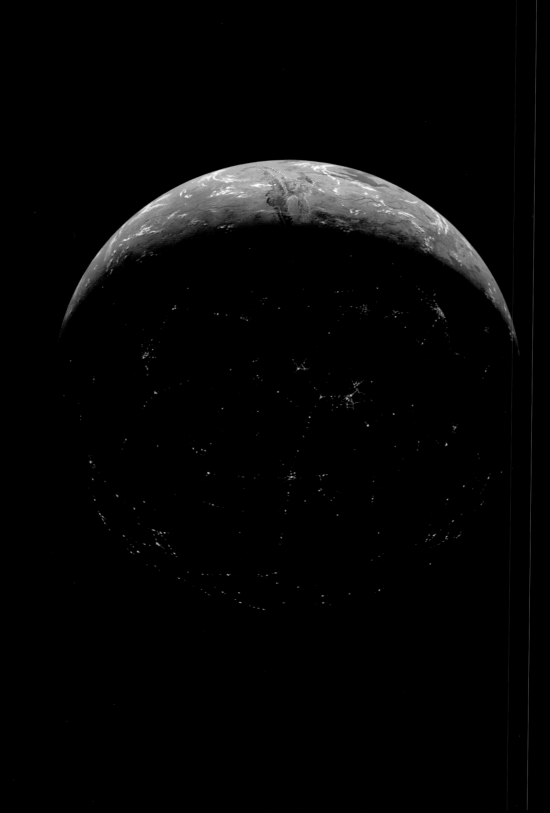

TERRAFORMING

Life isn't fragile; it's tough.
Here on Earth, life has endured heat, cold, and cataclysm,
has emerged from the sea to inhabit the land and the air,
and in human form has touched the Moon.
Someday we might carry the torch of life still farther outward,
to find a home on the planet Mars.

If Mars proves to be sterile,
we could make it suitable for life by transforming its environment,
through a global engineering approach called "terraforming."
Terraforming Mars would mean providing the red planet with life's three requisites—
water, energy, and organic molecules.
There's water there already;
human technology could provide the organics
and boost the amount of available energy.

In one plan, giant mirrors in space would warm the surface,
while genetically engineered plants and trees replenished the atmosphere.
Eventually the air would become breathable,
and the great-great-grandchildren of the first colonists might fish and farm
under the blue skies of a terraformed Mars.

Opposite: A terraformed Mars of the future is envisioned with a breathable atmosphere, shallow seas, and the lights of highways connecting new cities. Don Davis illustration.

We humans once imagined that we were at the center of it all.
Science has let the wind out of that vain claim.
Cosmic maps show that we live closer to the edge
than to the center of our galaxy,
and genetic maps show us occupying the tip of just one branch
of what's being called the shrub of life.

We're not at the center of the universe;
we're not at the top of the tree.
But then, neither is anybody else.
And, we're home.
We're part of the universe.

Yet, we're somehow also apart from the universe,
able to stand back and regard it objectively, as intelligent observers.
But what does it mean to be "intelligent,"
and how did intelligence originate here on Earth?

Are there intelligent beings out among the stars?
Will we ever communicate with them?
And what would we say?

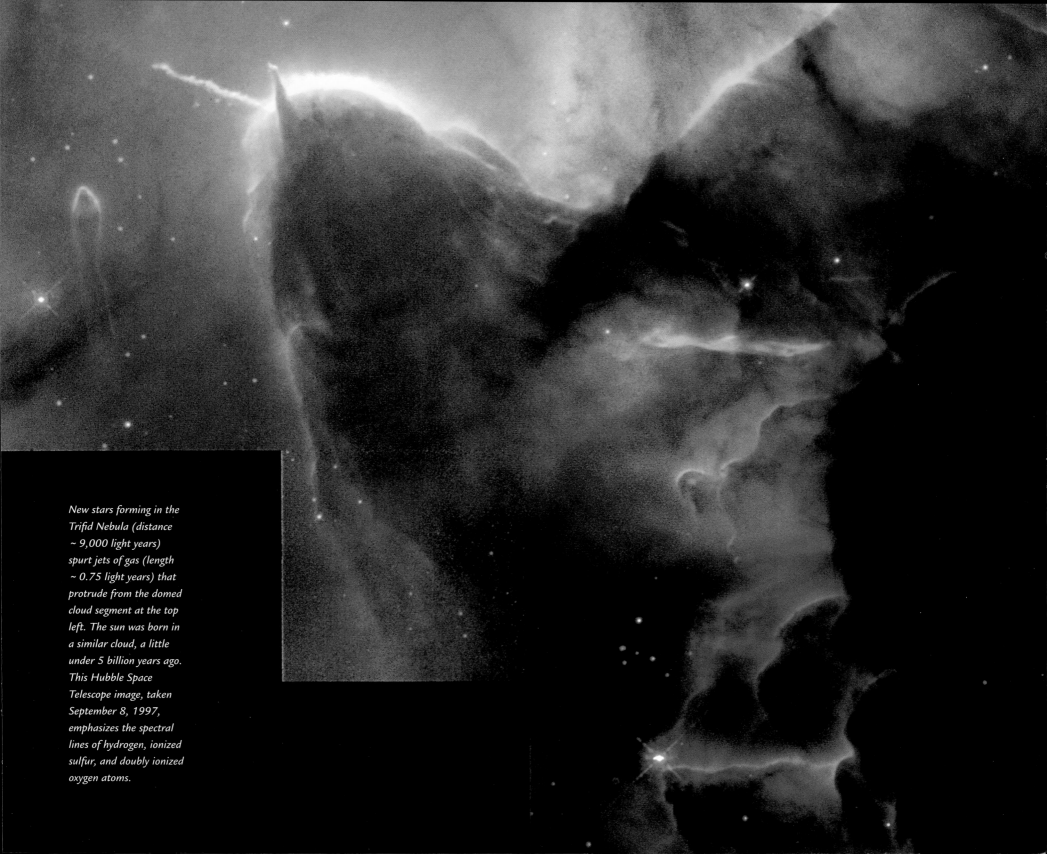

New stars forming in the Trifid Nebula (distance ~ 9,000 light years) spurt jets of gas (length ~ 0.75 light years) that protrude from the domed cloud segment at the top left. The sun was born in a similar cloud, a little under 5 billion years ago. This Hubble Space Telescope image, taken September 8, 1997, emphasizes the spectral lines of hydrogen, ionized sulfur, and doubly ionized oxygen atoms.

II.

IS ANYBODY LISTENING?

VISITORS

Ours is a restless species, and in our wanderlust
we have explored the globe and sent probes to look for life
on some of the Sun's other planets.
But we're also a communicative species,
and the way to search for life
on the trillions of planets circling other suns
may be through communication.
If there are intelligent beings out there,
who want to communicate.

If, someday, our descendants have explored the solar system
as our predecessors explored the Earth,
might they then venture to the stars,
and set foot on the soil of an extrasolar planet?
Can we sail through interstellar space
as mariners of old explored the islands of the sea?
Or is it better to communicate,
by listening for messages from space
and sending messages of our own?

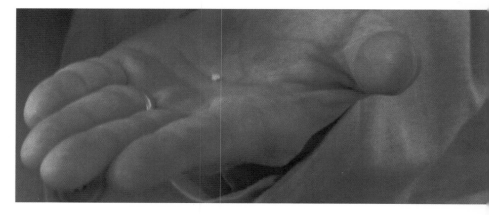

Any way you look at it, interstellar spaceflight is a daunting prospect.
It's pretty roomy out there.
If we could shrink the Sun to the size of a single grain of salt,
the Earth would be a microscopic grain,
orbiting at a distance of one inch,
and the orbit of the planet Mars would lie
within the palm of my hand.

The whole solar system,
the entire theater of wished-for human exploration
for centuries to come, could be encompassed
within the reach of my arms.

Yet even on this tiny scale, the very nearest star
would be more than four miles away—
beyond those distant mountains in the background.
A futuristic starship might make the trip one day,
if we could figure out how to pay its titanic fuel bill,
and how to fire up its engines without frying the Earth in its exhaust.
But interstellar spaceflight is likely to be expensive, and slow.

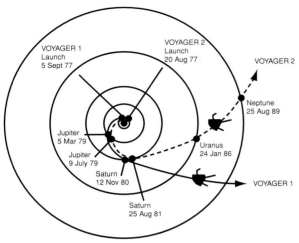

Left: Trajectory map shows the path taken by the two Voyager spacecraft out of the solar system. Voyager 1, the faster-moving of the two, was 11.5 billion kilometers from Earth on January 1, 2000.

Below: The Voyager phonograph record (diameter 30 cm), fashioned from bonded copper and encased in an aluminum canister, is expected to last for more than 1 billion years.

But if there are intelligent beings out there,
on the planets of other suns,
we may be able to communicate with them
even if they're too far away for us to visit them.

Our own boldest explorations remind us that the stars are far away.
The twin Voyager interstellar probes, launched in 1977,
surveyed the giant outer planets and are now leaving the solar system.
They carry a message—a gold-plated phonograph record
containing music and sounds of Earth,
for the benefit of any aliens who might one day intercept the space probes.
Although the Voyagers fly fifty times faster than a jet fighter,
they will take seventy thousand years to reach another star.

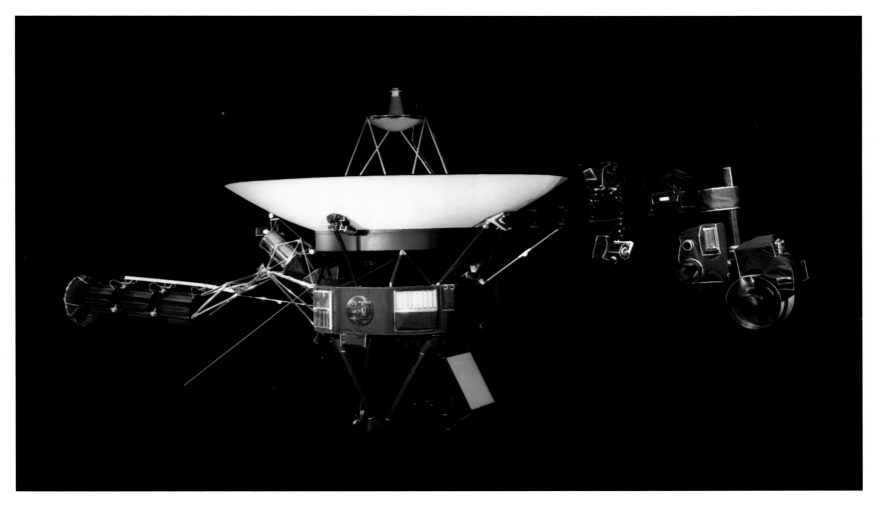

Full-sized model of the Voyager spacecraft shows the record mounted on its side. The probe's high-gain dish antenna is at top, imaging systems are on the near-side boom, and its plutonium power source is on the far boom.

Ever since the Renaissance, when Galileo established that other planets exist,
people have imagined that we would one day explore deep space.
The astronomer Johannes Kepler wrote to Galileo that
given the right "ships and the right sails for . . . space,
there will be men who [are not] afraid of the terrible distances."

131

Actually, the right ships already existed, in embryonic form.
Rockets began as a form of entertainment for the Chinese,
who turned them into weapons of defense against the Mongols.
The Italians turned them back into fireworks.

By the second half of the twentieth century,
rockets were being launched that were powerful enough to carry people aloft.
In a series of steps potentially as profound
as the migration of life from the seas onto dry land,
humans ventured into space.

Right: Launch of a Saturn 5 rocket, at 9:34 A.M. on July 26, 1971, carrying Apollo 15 astronauts David R. Scott, Alfred M. Worden, and James B. Irwin to the Moon.

Left: Stereographic images of fireworks over Ohkuchi Harbor, Matsusaka, Mie Prefecture, Japan, by Okuyuki (Yohji Maruyama), July 24, 1999.

The advent of spaceflight lent new immediacy to an old question:
If we can travel beyond our planet,
couldn't the denizens of other planets come here—
or be here already?

Anyone who spends a lot of time looking at the night sky
is likely, sooner or later, to see something he can't identify.
I've seen a few myself.
My most impressive sighting was of a V-shaped formation of white lights
that moved through the sky silently,
at what seemed to be an incredibly high speed.

Fortunately, I had a pair of binoculars around my neck,
so I was able to get a closer look.
And do you know what they turned out to be?
Migrating birds. A flock of migrating birds,
lights from a distant city lighting up their bellies
to make perfect little oval flying saucers.
If I hadn't had the binoculars handy,
it would have been a mysterious "UFO sighting."

Scanning the skies with binoculars from Rocky Hill Observatory, California, 1998.

The question isn't just what we see, but how we interpret it.
To claim that lights in the sky are alien invaders
is like claiming that they're angels:
It explains everything,
and, therefore, it explains nothing.
If we can hold off judgment until we have better information,
we can distinguish between dubious and sound propositions—
and between birds and spaceships.

When a real flying object unexpectedly shows up—
like the bright meteor that crossed the western part of North America
on the afternoon of August 10, 1972,
or the one that passed over New York state on a Friday night in the fall of 1992—
it typically results in much better pictures and more consistent eyewitness reports
than has any alleged UFO sighting.
Yet millions of people still believe that UFOs are alien spaceships.

Why?
Perhaps because we tend to project our deepest hopes and fears onto the sky—
imagining that the aliens are warlike and dangerous,
or that they are angelic benefactors
who have come to save us from ourselves.

Klaatu:

"Your choice is simple:
Join us and live in peace,
or pursue your present course and face obliteration.
We shall be waiting for your answer.
The decision rests with you."

THE INFOSPHERE

No one has come to rescue us, and it's unlikely that anyone will show up soon.
That fact may be telling us something—
that the best way to get in touch with intelligent extraterrestrials
is not exploration, but communication,
by sending, not starships,
but electronic signals across the interstellar seas.

The idea of communicating with extraterrestrials
has been in the air since the nineteenth century,
when scientists suggested that we signal the Martians
on a night when Mars passes near the Earth.
Their proposal was to dig a giant triangle in Siberia,
flood it with kerosene, and set the trenches afire.
The Martians would recognize the triangle
as a symbol of Euclidean geometry—
a sure sign of intelligent life on Earth.

Artist's conception of how the flaming triangle communications experiment, had it been carried out, might have looked if viewed through a telescope on Mars. Don Davis illustration.

In a modern version of the flaming triangle,
the Chinese artist Cai Guo-Qiang ignited
a ten-thousand-meter-long gunpowder fuse,
extended from one end of the Great Wall of China.
He called the installation "Project For Extraterrestrials."

Cai's work reinvents the partnership of fireworks and exploration.
But for sending real messages through space,
radio communication works better.
Radio can carry words, numbers, and pictures, cheaply and at the velocity of light.

Left: The pyrotechnic artist Cai Guo-Qiang's "Project For Extraterrestrials No. 10: Project to Extend the Great Wall of China by 10,000 Meters," employed 600 kg of gunpowder and 20,000 meters of fuse. Photo by Masanobu Moriyama, Jiayuguan City, China, 1993.

Near right: Marconi, in a news photograph taken June 16, 1922. The original caption describes him as "in his radio room trying to listen in on planet Mars."

Far right: Nikola Tesla (1857–1943) poses with a giant discharge coil at his Colorado Springs laboratory, 1899. Tesla made the photo, intended to impress his friends with his fearlessness, by combining more than 40 exposures of individual electrical sparks generated by the coil.

The almost miraculous speed with which radio carries messages
prompted even the earliest pioneers of radio technology
to imagine using this wonderful new device to communicate with other worlds.
Guglielmo Marconi, inventor of radio,
picked up what he thought might be radio pulses from space.
Nikola Tesla claimed that he, too, had received extraterrestrial signals.
Tesla generated powerful electrical bursts at his laboratory in Colorado Springs,
and fancied that the resulting radio noise could be received on other planets.

These imagined receptions were replaced by the real thing in 1931,
when a Bell Telephone Company engineer named Karl Jansky
accidentally detected the radio energy
emitted naturally by clouds of hydrogen gas in the Milky Way.
Grote Reber, an amateur radio enthusiast,
then built the world's first true radiotelescope—
a metal dish, thirty-one feet in diameter—
in the back yard of his home in Wheaton, Illinois,
and used it to map the Milky Way.

Today, radiotelescopes record the murmur of interstellar gas clouds,

the pulses of rapidly spinning neutron stars,

and the screams of colliding galaxies.

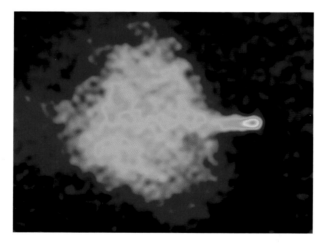

Radiotelescope images of (top to bottom) hydrogen gas in and around the galaxy M81, the "Duck" pulsar, a spinning neutron star that emits radio bursts; and the cannibal galaxy Centaurus A, with a plot of its radio emissions (the diagonal lobes from upper left to lower right) superimposed on a visual-light photograph (at center).

*Interacting galaxies
NGC2207 (left) and
IC2163 (right), imaged
with the Hubble Space
Telescope. Distance
across the frame is about
80,000 light years.*

Earth has been inadvertently broadcasting to the stars for decades.

The television premier of *Life Beyond Earth* was one such broadcast.

It reached the planet Saturn in an hour and a half, and continued into deep space.

A sphere of radio and TV broadcasts surrounds our planet.

Its radius is nearly one hundred light years, and is expanding at the speed of light.

Zooming out from the sphere of radio and television broadcasts from Earth reveals how little of our galaxy these inadvertent signals have yet reached. Don Davis illustration.

A thousand stars lie within the sphere
of radio and television broadcasts
that have leaked into space from Earth.
If there are any listeners out there,
they could, in theory, intercept these broadcasts.
The farther they are from us, the older the news they're receiving.

Alpha Centauri is a star only four light years away,
so Earth's broadcasts reaching Alpha Centauri today
are only four years old.
The star Capella is forty light years away.
If anyone's listening there, what they're hearing from Earth
are the hits of the fifties.
The golden age of jazz is just dawning at Beta Pictoris,
a star that shows evidence of having planets forming around it.
Any radio buffs at Mizar, a star in the handle of the Big Dipper,
are in a position, just about now, to be picking up Earth's very first broadcasts.

Radios from the early twentieth century (foreground), the forties (middle), and the nineties (background) are arrayed to demonstrate that the farther from Earth extraterrestrial listeners are, the older are the broadcasts they would be hearing today.

One of the twenty-seven dish antennas (diameter 25m) of the Very Large Array, a radiotelescope on the plains of San Agustin near Sorocco, New Mexico. The VLA has the sensitivity of a single dish 130m in diameter and the resolution of an antenna with a diameter of 36km.

We're leaking radio messages into space,
and the inhabitants of other planets, if they exist,
might be doing the same thing. Large radiotelescopes on Earth
could detect such radio leaks from civilizations in nearby star systems,
as well as stronger signals dispatched from planets thousands of light years away.
This realization is the keystone of SETI—the Search for Extraterrestrial Intelligence.

Some scientists doubt that there are any intelligent beings out there.
One of their arguments is called "Fermi's Question."
It's named for the Italian physicist Enrico Fermi, who asked,
"Where are they?"
His point was that technologically advanced aliens, if they exist,
should have visited Earth already.
Since they evidently have not, they don't exist.

I performed an experiment to test the validity of Fermi's question.
At home alone one night, I decided to have lobster for dinner.
So I set a place, opened the door to the street, and waited for a lobster
to show up and crawl onto my plate.

Hours passed.
At eleven p.m.
I ended the experiment.
No lobster had appeared.
So, I concluded, there are no lobsters on Earth.
Since we know that lobsters do exist,
clearly there was something wrong with my reasoning.
The error was, of course, that I'd failed to take the lobsters' preferences into accoun
Lobsters have their own agenda. They don't want to come to my house.
But the fact that they don't show up doesn't mean they don't exist.
As the SETI scientists like to say,
"Absence of evidence isn't evidence of absence."

Waiting for a lobster to appear. Does the fact that extraterrestrials haven't landed on the White House lawn mean they don't exist? Photographed in Florence, Italy, by Bob Elfstrom.

A volume of the imaginary Circuli Lactei Planetae ("Planets of the Milky Way"). To cover the planets estimated to exist in our galaxy, at one per page, would require about a billion such volumes, more than could be stored in all the university libraries on Earth.

Suppose we had an atlas that devoted but one page
to each planet in the Milky Way galaxy.
That's a lot of planets.
To page just once through all its volumes,
glancing at one planet per second, day and night,
would take over ten thousand years.
As the historian Thomas Carlyle said,
contemplating the prospect of billions of worlds,
"If they be inhabited, what a scope for misery and folly.
If they not be inhabited, what a waste of space."

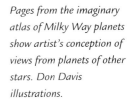

Pages from the imaginary
atlas of Milky Way planets
show artist's conception of
views from planets of other
stars. Don Davis
illustrations.

At astronomical observatories like this one,
on the island of Hawaii,
astronomers are finding evidence of planets orbiting nearby stars.

Astronomer Paul Butler:

"We only want to know two things:
What fraction of stars have planets,
and what fraction of those planetary systems are similar to our own solar system?
It wasn't until the early part of this century that it became technically feasible
to be able to find planets around other stars."

The twin domes of the Keck observatory contain identical telescopes, each with a 10-m light-gathering mirror. Linking the two electronically and aiming them at the same object produces an interferometer that can achieve the resolution of a telescope with a mirror as large in diameter as the 85-m distance separating them.

Astronomer Geoffrey Marcy:

"Our technique involves watching a star
to see how it is gravitationally pulled by its attendant planets.
An Earthlike planet is simply too small to yank around its host star.
But a Jupiter, being some three hundred times more massive than an Earthlike planet,
does have enough gravitational *oomph* to yank its star around.
And so for the moment, all we can detect are Jupiters and maybe Saturnlike planets.
We watch the star night after night, month after month,
and if the star does wobble, it must be doing so because a planet is pulling on it.
It's a bit like a dog owner having a little poodle on the end of a leash.
The poodle, being very low-mass, can still yank its owner around,
if the poodle goes around in circles.
In this case, the low-mass planets can yank around the star.
In the case of a poodle, the leash is a leather tether.
In the case of the planets, the leash is gravity.
We look for very teeny shifts in the spectrum that tell us
that the star is wobbling in space—
moving alternately toward us and away from us.
If the star's moving toward you, its entire spectrum will be shifted to the blue,
and if the star's moving away from you, then the star will be shifted to the red."

Future telescopes operating in space
could be linked together to create the equivalent of a lens larger than the Earth.
Such orbiting observatories might be able to detect signs of life
on the planets of nearby stars.

Artist's conception of a space interferometry mission: Twin telescopes, free from the distorting effects of Earth's atmosphere, produce high-resolution images of distant galaxies.

Telescopes have incited speculation about extraterrestrial life
ever since the summer nights, four hundred years ago,
when Galileo first trained a telescope on the sky.
Galileo called the universe a "great book . . .
written in the characters [of] geometrical figures."
This was a revolutionary idea. The books that mattered in Galileo's day
were the ones written by ancient authorities,
like Aristotle and Ptolemy.

They said that the universe was a kind of nutshell with the Earth at its center,
and that the Sun and planets were ethereal wafers pasted on the inside of the shell.
In that kind of universe, there could be no life on other worlds
because there *were* no other worlds.
The stars and planets were like the decorations of the insides of the great cathedrals—
beautiful, pristine, nearby, and almost perfectly useless.

But when Galileo turned a telescope on the sky,
he found that the planets and the moon didn't look like wafers.
They looked like rotund worlds, resembling the Earth in some ways.
And when he looked at the Sun,
he found that it wasn't perfect, but was besmirched by sunspots.
As Galileo observed the Sun, in Florence,
week after week, he could see the sunspots,
which are giant magnetic storms on the surface of the Sun,
moving across the solar disk—
not the behavior of a flat, static object, but, as we would say today,
of a rotating ball of plasma, a *star*.

Galileo demonstrating his telescope to the Doge and senators of Venice. Fresco by L. Sabatelli, 1841.

159

Galileo's observations blew the roof off the universe.
Aristotle to the contrary, the Earth is not the only world,
housed in a starry nutshell.
It's one among many worlds,
and there's no nutshell at all.
The Sun is a star, and every star is a sun.
The mathematical laws of nature that pertain here on Earth
work everywhere else as well. And if there's life here, why not out there?

Left: Woodcut from L'Astronomie, *by the astronomer Camille Flammarion (1842–1925).*

Opposite: Fontenelle Meditates on the Plurality of Worlds, *1791, portrays the popular French author Bernard de Fontenelle, who noted that "the earth swarms with inhabitants," and asked, "Why then should nature, which is fruitful to an excess here, be so very barren in the rest of the planets?"*

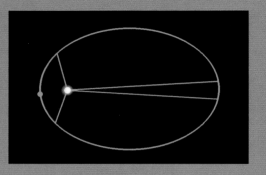

Galileo's experiments touched off a scientific revolution
that awakened humankind to the possibility of life beyond Earth.
In Germany, Johannes Kepler identified the laws that govern the motions of the planets.
Kepler found that planets move faster when close to the sun
and slower when farther away,
in just such a way that their orbits always sweep out equal areas in equal times.
This is a genuinely universal law, true of any object in orbit—
from the moons of Jupiter to the planets of distant stars.

Above: A planet or other body moving in an elliptical orbit moves faster when close to its sun (left) and slower when far away (right). Kepler found that any two such triangles, describing equal intervals of time (e.g., one month) in orbit, have equal areas. Don Davis illustration.

Right: Isaac Newton (1642–1727) in middle age, painted by Sir Godfrey Kneller, 1689.

Opposite: Hand-tinted portrait of Albert Einstein (1879–1955)

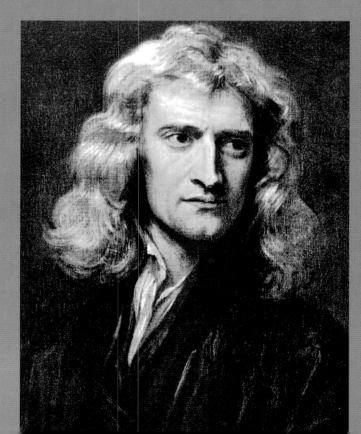

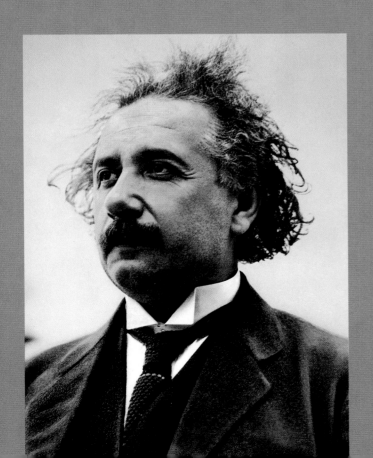

In England, Isaac Newton showed
that Kepler's laws result from the force of gravitation,
exerted by the Earth and every other massive object.
It is by using Newton's equations that today's space navigators
are able to put probes into orbit around Venus, Mars, and Jupiter.

In the twentieth century, Albert Einstein
cleared up disparities in Newton's theory
by composing a broader account of gravitation,
the general theory of relativity.

Relativity implied that cosmic space is expanding.
The idea seemed so outlandish that Einstein himself rejected it.
But big new telescopes were being built,
and astronomers using them were able to identify
individual stars in other galaxies,
and determine that they obey the same physical laws
that pertain here on Earth.

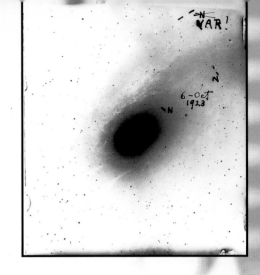

Edwin Hubble (1889–1953) examining the spectrum of starlight from another galaxy. Shifts in spectral lines showed that galaxies generally are moving apart from one another as cosmic space expands.

Above: Photographic plate taken October 6, 1923, on which Hubble excitedly scrawled the abbreviation "VAR!," indicating that he had identified a Cepheid variable star in the Andromeda galaxy. Such stars, employed as distance indicators, made it possible to measure the rate at which the universe is expanding.

Opposite: The Andromeda galaxy (M31, distance ~2 million light years, diameter ~100,000 light years) is the closest major spiral galaxy to the Milky Way. Hence Andromeda provides a splendid view of a galaxy similar to ours— as would ours to astronomers living there.

In 1929, the American astronomer Edwin Hubble,
without knowing that relativity implied that the universe expands,
discovered that the distant galaxies are indeed moving apart from one another,
just as Einstein's theory had predicted.
Four centuries after Galileo, humans had learned
that the universe is all of a piece, expanding and evolving
in accordance with one set of physical laws.

Living creatures are made of the same stuff
as are the stars and planets,
and they obey the same laws.
And so, in theory, life could exist
all over the universe.

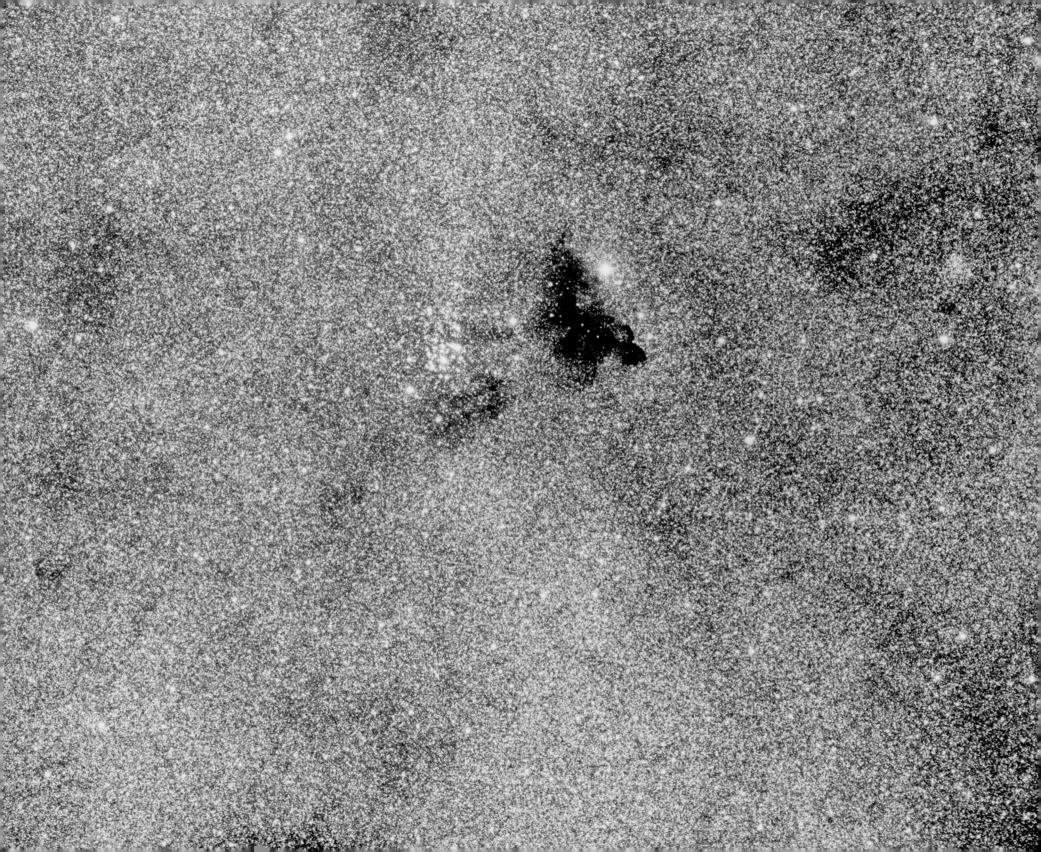

MESSAGE IN A BOTTLE

**There are more stars in the universe
than there are grains of sand at the seashore.**

*Rich star field in
Sagittarius, looking toward
the center of our galaxy,
shows older yellow stars
and young blue stars like
those in the cluster near
center (NGC 6520) that
are still entangled in the
cloud of dust and gas from
which they formed.
Photographed by David
Malin, July 16, 1980.*

Our human desire to communicate runs so deep
that we're willing to send messages
even though we're not sure who might receive them.

*Schoolchildren (Makana
Phillips-Parker, Anastasia
Sumrall, and Edward Ho)
from Waimea and Waikola,
Hawaii, compose messages
to extraterrestrials, seal
them in bottles, and
dispatch them on the
Pacific tide.*

"Dear Readers, what is life like on your planet?"

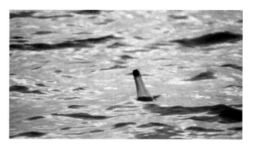

"Do you have wars?"

Today, we have the means to send messages
to the planets of other stars.
But is anyone listening?

"Do you have any special abilities, like zapping people?"

The 70-m radiotelescope of the Goldstone Deep Space Communications Complex in the Mojave Desert, California.

If we are to communicate with the inhabitants of other worlds,
there must be creatures out there who have language, radio technology,
and the ability to keep transmitting for centuries at a time,
so that their signals can cross hundreds of light years
of interstellar space.

Yet we don't know how *we* learned to use language,
or how long *we* may be willing and able to listen
for messages from the stars.

If we understood how intelligence arose on Earth,
we'd have a better idea of how often intelligence arises on other planets.
The size of our ancestors' brains more than doubled
during the past two million years,
but nobody knows quite why.

Somewhere along the line, about fifty thousand years ago,
human speech advanced from grunts and warning cries
to the use of words as abstractions,
capable of probing beyond the realm of immediate experience.

In much the same way that prehistoric painters
brought the daylit world into the darkness of caves,
ancient poets painted word pictures
of events remote in space and time.

*Cave fresco (size >6m²) in
Rhone-Alps region, France,
has been carbon-14 dated
at 31,000 years of age.
Some black accents were
added 4,000 years later by
torch-snuffing technicians.*

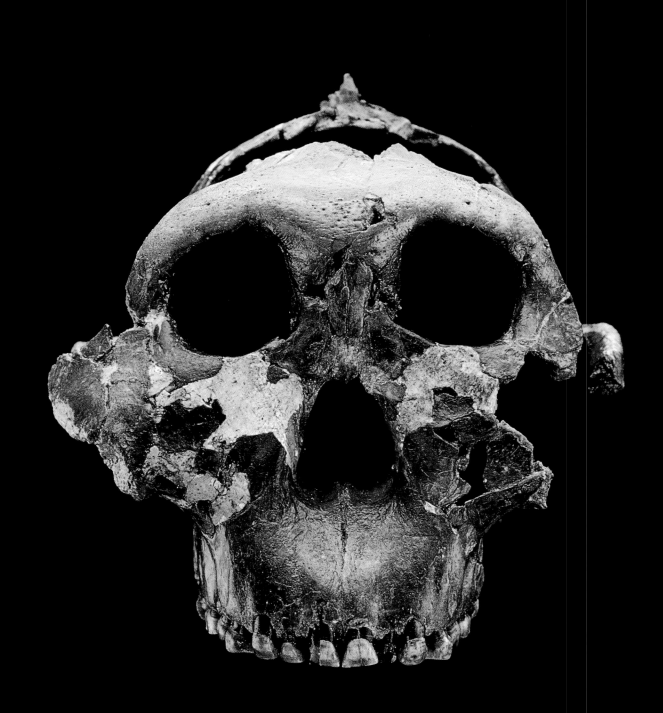

The skulls of an Australopithecus boisei (left, circa 2.3–1.4 million years ago; cranial capacity ~500 cc) and Homo sapiens (right, circa 200,000–100,000 years ago to present; cranial capacity >1,000cc) come from opposite sides of the watershed in time when, for reasons not at all well understood, our ancestors first formulated language. Photographed by David Brill in Tanzania (left) and at Harvard University (right).

From that point on, evolution began to select,
not just for brute force and endurance,
but for abstract intelligence as well.
Nobody knows just when this happened, or how.
Until we do, it's going to be difficult to estimate
how often intelligence has arisen on other planets.

Left: Egyptian hieroglyphs, Middle Kingdom (2065–1785 B.C.), from a limestone bas-relief near the White Chapel of Karnak. Dedicated to the fertility god Amon-Min.

Right: Sumerian inscription on limestone tablet (circa 4000 B.C.; lower Mesopotamia; proto-urban phase) listing proper names. The hand at the upper left records the name of a landowner.

Once our ancestors could make sentences,
they could make plans.
In hunting and warfare, the balance began to shift
from strength to strategy,
with strategy tending to win out in the end.
As Abraham Lincoln put it,
"Force is all-conquering, but its victories are short-lived."

Language is the key to communication.
Symbolic language enables us to do mathematics,
describe events that happened far away,
and make a lasting record of our thoughts and feelings.

If we are to communicate with extraterrestrials,
they, too, must be able to manipulate symbols.
To search for intelligent life in the universe is, therefore,
to search for another species similarly endowed
with an aptitude for symbolic language.

If they don't have language, they may be smart,
but we're not apt to hear from them.
Creatures lacking symbolic language
cannot send complex messages through space
or much of anywhere else. They can cry out in alarm
or coo tenderly to their young,
but they can't tell someone far away about their planet,
their history, or their science. Indeed, without language
they can't *have* any real history or science.

In that sense, looking for intelligent extraterrestrials
means looking for beings rather like ourselves.
Not that they'd resemble us physically,
or speak in a human tongue.
But they would, by definition, have language.

And where there is language,
there is hope for translation, and, in translation,
an end to the long twilight of human loneliness—
during which we have known the delights and torments
of being uniquely thoughtful, communicative,
and self-aware.

Ancient Chinese ideogram
meaning "to write."

Is intelligent life merely an accident?
Popular accounts of science often say so,
claiming that life and intelligence are trivial.
One scientist called life "a fancy kind of rust,
afflicting the surfaces of certain lukewarm, minor planets."
Another portrayed humans as "scum,"
a cosmic afterthought,
clinging to a small planet
in an uncaring universe.

Above: Time exposure photograph of the Dallas, Texas, skyline, February 1999, by Jeremy Woodhouse on a 6 x 17 cm negative.

Near right: A flock of lesser flamingos (Phoenicopterus minor), Kenya, 1996.

Far right: School of fish in a marine cave, Cocos Islands (Pacific Ocean, 900 km northeast of the Galapagos), May 1993.

But there's a new vision on the horizon,
one that sees intelligence as an "emergent property,"
apt to appear on any planet
that has a sufficiently complex biological system.

The patterns formed by birds in flight
and by schools of fish
are emergent properties.
They can be seen, not in individual fish or birds,
but only in large assemblages of them.
According to this theory, emergent properties can be understood
only by studying the system as a whole.

The spiral arms of galaxies are fireworks shows
created when brilliant young stars are formed
by density waves moving across the galactic disk.
Millions of stars must be born
for the spiral pattern to appear.

*Left: Spiral galaxy NGC
4603 (distance ~108
million light years), imaged
by the Hubble Space
Telescope, 1999.*

*Opposite: Spiral galaxy
NGC 1232 (distance
~ 100 million light years)
photographed on September
21, 1998, with the
European Southern
Observatory's Very Large
Telescope in Chile.*

Negative prints highlight spiral structure in galaxies. Clockwise from upper left, this page: NGC 4321, NGC 7412, NGC 4254 (in blue light), NGC 5247 Opposite page: NGC 4254 (in red light), NGC 5364, NGC 628, NGC 5457. Photographs by Allan Sandage.

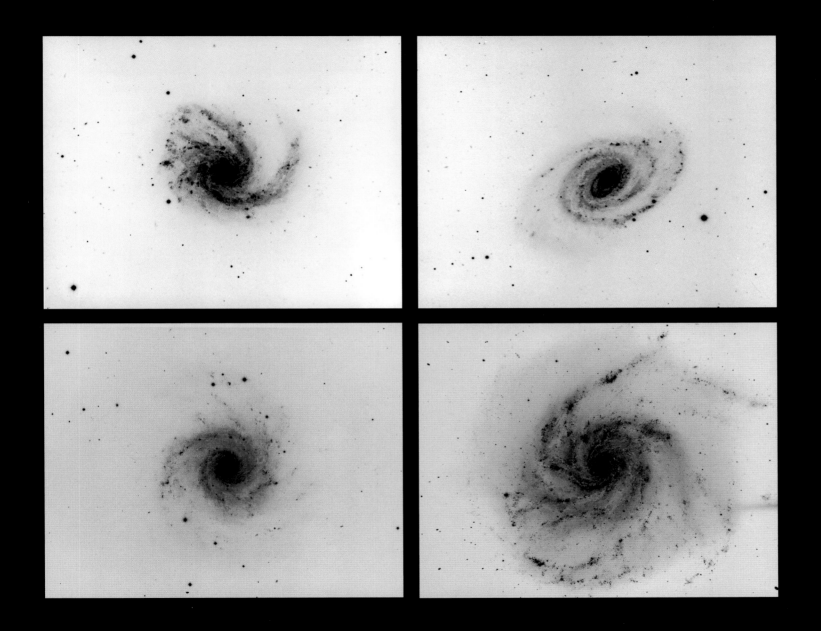

Weather is an emergent property.
A low-pressure front is made of molecules of air and water,
but we cannot perceive it just by looking
at individual molecules.
Only when we consider the system at large
does it emerge as a thunderstorm.

If intelligence emerges normally
in the biospheres of living planets,
as naturally as weather patterns emerge
in their atmospheres, there may be
many thinking beings on other worlds.

Science traditionally has concentrated on the parts—
the approach known as reductionism.
But if intelligence is an emergent property,
arising as naturally as clouds gathering to make a storm,
then thought may not be incidental to the workings of the universe,
but, in some sense, the point of it all—
the central theme in a cosmic symphony.

Left: Cloud-to-cloud lightning in a thunderstorm, Sonora Desert, southern Arizona, August 1998. Photo by Will Milan.

Opposite: Slow-moving thunderstorm cell (distance ~40 km) over western Sydney, Australia, photographed by Michael Bath, November 27, 1994.

DESTINY

There's an Islamic saying that three things are known to no man—
the hour of his death, the true name of Allah,
and the source of his next meal.
In the Jewish tradition there's the story of a rabbi
who's stopped after curfew by a Cossack who says
"Where are you going, old man?"
and answers, "I don't know."
The Cossack arrests him, and as the door to his cell slams shut
the rabbi says, "You see? I told you I didn't know where I was going."

Which is pretty much our situation as members of an intelligent species
who don't know if there are any other intelligent beings
out there in the universe.
We think it's good to be smart,
that it bodes well for our destiny,
and should for other species, too, in other worlds.
But we don't really know where we're going—do we?

Does the fact that humans are intelligent mean that we're likely to survive?
Or will we blow ourselves up, or pollute the Earth so badly
that we can no longer live here?

The physicist Richard Gott has devised a formula
for predicting how long the human species is going to last.
He begins with the assumption that there's a 95 percent likelihood
that we're living during the middle 95 percent
of humanity's total tenure.
To date, Homo sapiens have existed for about two hundred thousand years.

*Cosmological model
portrayed by the Syrian
theoretical astronomer Ibn
ash-Shatir, 1367.*

Richard Gott:

"There's a 95 percent chance that you're located in the middle 95 percent of human history, and that, therefore, the future longevity of the human race is likely to be longer than 150 years, but less than 7.8 million years.
Now, those numbers are interesting,
because they give us a total longevity
that's quite similar to that of other species."

All beings who have learned that the universe is old
must also know that, by comparison, *they* are young.
The longer alien civilizations typically last,
the better our chances of communicating with them.

The lights on this tree represent technological civilizations
that might have arisen in a given galaxy over a period of one million years.
The length of time that each light stays on
represents the period during which each civilization
is willing and able to make radio contact
with the inhabitants of other such worlds.

If the communicative worlds typically stay on the air for five hundred thousand years,
their persistence is rewarded: Each finds that it has plenty of contemporaries
with whom to make radio contact.
So if civilizations last a long time, SETI
—the Search for Extraterrestrial Intelligence—
is a cinch.

But watch what happens
if alien civilizations typically last only one hundred thousand years
before they become extinct, or lose interest.
Now, they find that there are only a few worlds
on the air at the same time they are.

Does the fact that humans are intelligent mean that we're likely to survive?
Or will we blow ourselves up, or pollute the Earth so badly
that we can no longer live here?

The physicist Richard Gott has devised a formula
for predicting how long the human species is going to last.
He begins with the assumption that there's a 95 percent
likelihood
that we're living during the middle 95 percent
of humanity's total tenure.
To date, Homo sapiens have existed for about two hundred thousand years.

Christmas tree lights illustrate how many civilizations in a given part of a galaxy would be able to communicate with one another at a given time, based on how long they typically remain viable. The tree opposite, top, depicts the situation if civilizations last an average of 500 million years: Many coexist and can communicate. But if their average viable lifetime is 100,000 years (opposite, bottom) they have fewer companions, and if they last only 100 years (above), the pickings are slim.

What monuments might they have built,
to leave some trace of themselves in the annals of cosmic history?

Perhaps they long ago established a permanent network
to link inhabited planets and preserve a record of their histories.
If so, the first signal we receive could come from an interstellar Internet.

Deployed over eons by robotic spacecraft,
such a network could bring libraries of information
within relatively close reach of emerging worlds like ours.

Communication can bridge time as well as space,
revealing the histories of societies that disappeared long ago,
and offering clues to how our species might best navigate its way
toward the dim and distant future.

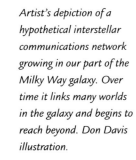

Artist's depiction of a hypothetical interstellar communications network growing in our part of the Milky Way galaxy. Over time it links many worlds in the galaxy and begins to reach beyond. Don Davis illustration.

We don't know how many alien civilizations exist,
how long they typically last,
or how many of them take the trouble to broadcast signals.
What if everybody's listening, and nobody's sending?

Scientists involved in SETI seldom transmit anything—
either because they'd have to wait decades or centuries to get an answer,
or out of fear of betraying our presence to aliens who might,
for all we know, be hostile toward us.

But what if someday we did want to send a message,
a powerful message beamed to a planet
where we had some reason to expect
that there was somebody listening?
What would we say?

Overleaf: Milky Way,
photographed by (left)
John Gleason and (right)
John Gleason and Steve
Mandel.

"I would want to know if they're bad guys or good guys.
If they're good,
we can work with them, right?"

—Astrophysicist France Cordova

"'Dear Extraterrestrial:
Please respond to this message by sending a radio signal
back on the frequency at which you receive this message,
making due allowance, of course, for your motions around your star.
Send that signal at a power level such that it arrives on Earth
with at least ten to the minus twenty-three watts per square meter,
and leave your transmitter running long enough,
let's say a few months,
so that one of the searches now operating on Earth will receive it.'
Then we can really get a conversation going."

—Physicist Paul Horowitz

"I would say, 'Please forgive us for making so much noise.'
Talking is what we do all the time,
and maybe the reason we don't hear the aliens
is because they are different, and they keep quiet."

—Physicist Freeman Dyson

"'What an awesome, self-unfolding,
miraculously generative universe
we are fortunate enough to share.'
That's what I would say."

—Complexity scientist Stuart Kauffman

"What I want to know from you is, what's your biochemistry?
I really want to know whether life has to be done through DNA
or whether there are lots of other ways,
or I need another experiment to find out.
And the other thing I will give to you is the *B Minor Mass*,
because that's the best thing we've ever done.
But I'd like to know whether you've ever done anything that beautiful,
and if so, what was it, and share it with us."

—Paleontologist Stephen Jay Gould

"Our intelligent species has been around for two hundred thousand years.
And we have sent members of our species to our neighboring moon,
and we've learned something of the laws of physics,
and we've learned where we are in the universe.
How are you doing?"

—Physicist J. Richard Gott III

Timothy Ferris:

"Greetings from one of the species that inhabit Earth,
a blue planet orbiting an average star.
We call ourselves *Homo sapiens,* by which we mean
that we are capable of mathematics, science,
and the technology by which this message is being sent.
We're the only such creatures on this world, and, so far as we know,
the only ones to have lived here throughout its long history.

"Science has brought a better life for millions of us,
but many more of us still live in poverty, tyranny, and ignorance.
Sometimes we wonder if there's intelligent life on Earth.

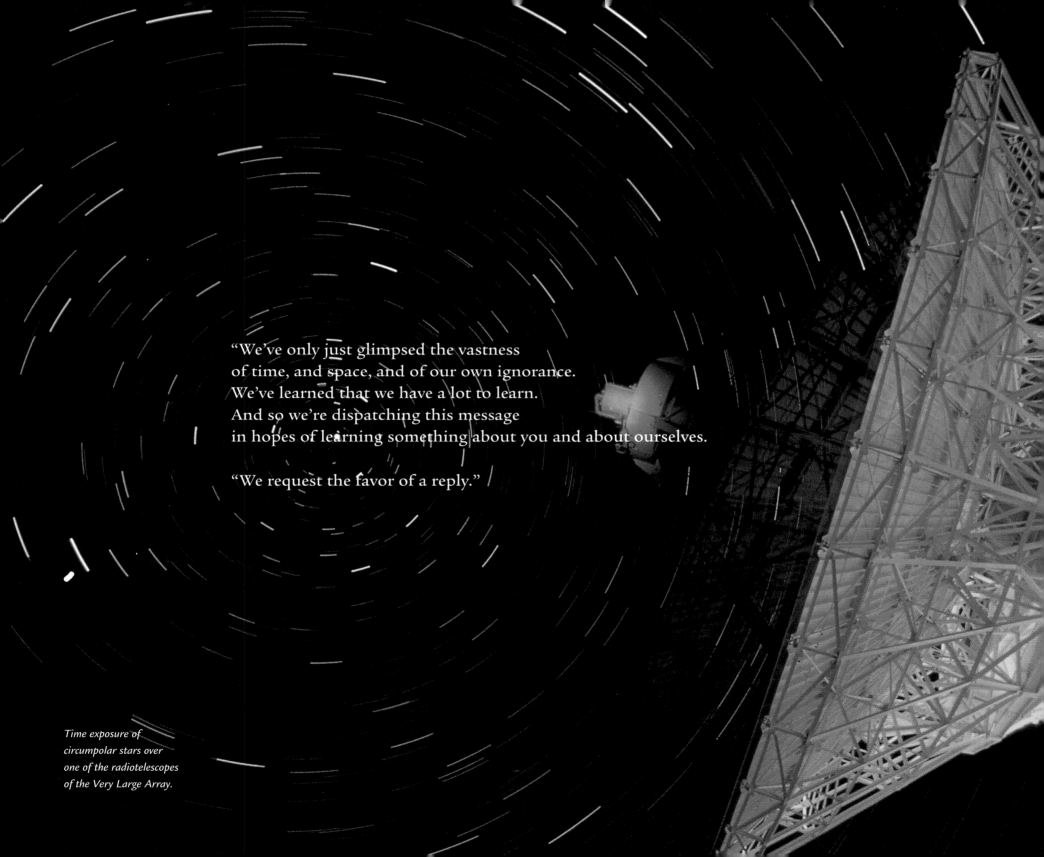

"We've only just glimpsed the vastness
of time, and space, and of our own ignorance.
We've learned that we have a lot to learn.
And so we're dispatching this message
in hopes of learning something about you and about ourselves.

"We request the favor of a reply."

*Time exposure of
circumpolar stars over
one of the radiotelescopes
of the Very Large Array.*

OTHER VOICES

Astrophysicist France Cordova:

"Ever since I was very little, I wondered about how I got here,
and how the universe got started.
I used to love to go out and look at the stars at night,
and as I came to know much more about the universe,
it was just inconceivable to me that we were all alone
among all those stars and all those galaxies."

Physicist Paul Horowitz:

"We can't prove, of course, whether there's life beyond Earth,
because we haven't found it yet.
But what happened, on a probably ordinary planet, circling a certainly ordinary star,
and happened quickly—namely the beginnings of life, 3.8 billion years ago—
must have happened in many other sites in the universe.

"Can we be the only ones?
Implausible. In fact, in my humble opinion, impossible.
Four hundred billion stars in our galaxy alone.
Many of them undoubtedly have planetary systems,
as we're just beginning to discover.
Fifty billion other galaxies.
What are the chances in 450 billion stars, 50 billion galaxies, one,
one intelligent civilization? And guess what? We're it.
Completely impossible."

Biologist Norm Pace:

"Old notions of evolution
had it that there was a ladder of evolution,
proceeding from the simplest organisms to the most complex.
But that really doesn't hold water.
We as complex organisms would not do very well
in places where simple organisms do very well—
for example, in the crust of the Earth,
or around submarine hydrothermal vents.
All organisms that are alive today are exquisitely fitted
to the particular environment that they occupy.
And one organism in a particular environment is not very good, necessarily,
in another environment.
Life isn't a ladder.
Life is a selection
for wherever that organism happens to be."

Neuroscientist Gerald Edelman:

"We don't know whether life exists out there,
but if it does, I think it's a fair guess
that it would have to follow the principles of Darwinian evolution.
Let's say that somehow there was a creation out there
of circumstances that gave rise to life.
It certainly would not survive under a changing environment
unless there was something like natural selection taking place."

Complexity scientist Stuart Kauffman:

"Before language, even, our artifacts become a shared means
by which we could accumulate knowledge of how to make our way in the world.
I think that natural evolution of know-how
in the coevolving population of critters
is probably going to be more or less expectable.
Whether that leads to consciousness or language is quite another issue."

Paleontologist Stephen Jay Gould:

"The main reason for considering intelligence as accidental
is pretty clear: It's only evolved in one species
after four and a half billion years of the history of the Earth,
which is about half the Earth's potential history,
if the Sun's due to explode in five billion years or so.
That's pretty amazing.
There are only four thousand species of mammals; we're a minor group.
Consciousness has only arisen in one species, us,
of a minor order of mammals, the primates,
with fewer than two hundred species in toto.
You have a million named species, including about five hundred thousand beetles.
If intelligence was such a good thing,
and it was so obviously of Darwinian benefit,
and it was an easy thing to achieve,
I assume other lineages would have it, and they haven't.
And yet they're doing very well."

Physicist J. Richard Gott III:

"Albert Einstein was very smart,
but he didn't live orders of magnitude longer than other people.
And what we've learned is that smarter species
don't tend to live longer than other ones.
This is an unfortunate thing to notice."

Physicist Freeman Dyson:

"Species, generally speaking, last a few million years.
That's sort of typical of a species.
And if that's true of our species,
we're in good shape for another million years or so.
That would be plenty of time for most people.
However, the whole idea of species is fading away.
As soon as you have genetic engineering—
genes being transferred from one species to another,
which is already happening—
we could become many different species, if we are species at all,
in a time that's much shorter than two million years.
There's just a huge difference between a hundred years and a million years.
In a hundred years we'll be the same as we are,
but in a million years, probably not.

"So we can imagine that over billions of years, intelligence
has grown and developed far beyond any sort of intelligence we now have,
and that it might, in fact, become a major player in the physical development of the universe—
reorganizing galaxies to move immense quantities of matter and radiation
from one part of the universe to another. We might, in fact, become creators
in producing a universe in which we can live forever.
That's at least a dream which is not altogether pointless to think about."

Distant galaxies crowd a tiny piece of sky, about 1 percent the diameter of the full Moon, in this "deep-field" image, assembled from 276 separate exposures made with the Hubble Space Telescope, December 18 to 28, 1995. Light from the most remote of these galaxies had been traveling through space for some 10 billion years when it reached Earth.

UNITS OF MEASUREMENT

SYMBOLS

Velocities are expressed in abbreviations,
e.g., km/sec for kilometers per second, km/hr for kilometers per hour

= means equal

~ means approximately equal, e.g., the velocity of light = 299,792 km/sec ~ 300,000 km/sec

> means more than; e.g., 3 > 1.

< means less than; e.g., 1 < 3.

METRIC AND ASTRONOMICAL UNITS

1 millimeter (mm) = 0.04 inch

1 centimeter (cm) = 10 mm

1 meter (m) = 1,000 mm = 1.09 yard

1 kilometer (km) = 1,000 m = 0.62 miles

1 light year = ~ 6,000,000,000,000 (6 trillion) miles

TEMPERATURES

In the Fahrenheit (F) scale, water freezes at 32° and boils at 212°
In the Celsius (C) scale, water freezes at 0° and boils at 100°
In the Kelvin (K) scale, water freezes at 273° and
boils at 373°

$$F = 9/5 \, C + 32$$
$$F = 9/5(K - 273) + 32$$
$$C = 5/9(F - 32)$$

ACKNOWLEDGMENTS

Life Beyond Earth is based on the film of the same name, which involved the work of a great many individuals, among them Series Producer Linda Feferman, Executive Producer Elizabeth Brock, and Producer Jamie Stobie, as well as Ron Devillier, Brian Donegan, Greg Diefenbach, and Ciara Byrne of Devillier Donegan Enterprises and Mary Jane McKinven of the Public Broadcasting System. Principal cinematography was by Jon Else and Bob Elfstrom. *Life Beyond Earth* was based on an original idea by Steve Most, and contains additional writing by Steve Most; complete screen credits appear in the film. Special thanks for their help with the book go to my assistant, Terra Weikel; photo researcher Clare Rickinson; Randy Brinson, Cai Guo-Qiang, Lisa Day, Don Davis, Jill Freidberg, Andy Fraknoi, David Grinspoon, Donald Johanson, Monica Lamontagne, Zoltan G. Levay, Doug Lundberg, and Andy Perala; and the staffs of PBS and KCTS-TV.

And, as always, thanks, Cal.

PHOTO CREDITS

The author gratefully acknowledges the individuals and institutions that provided these images, and invites corrections of any errors that may have occurred despite careful photo research. Abbreviations are defined at the end of the credit list.

Page	Credit
12–13	John Gleason/Celestial Images
18	ESO
20	LBE
21	Courtesy of Hubble Heritage Team (AURA/STScI/NASA)
22 (left)	SOHO
22 (right)	Robert Gendler
23	John Gleason/Celestial Images
24	Biophoto.associates/Photo Researchers, Inc.
25	© 2000 Keith Clark
26	NSSDC
27	Photography NASA; digital scan and image © 1999 Michael Light Studio
28	JPL
29	Tony and Daphne Hallas/Astro Photo
31	UKS
32	Frans Lanting/Minden Pictures
33	© 1987/Jim Brandenburg/Minden Pictures
34, 35	AAO
36, 38, 39, 40 (left, right), 41, 42	LBE
43	JSC
44, 45	CORBIS/Bettmann
47, 48, 49	The Natural History Museum, London
50, 51	CORBIS/Bettmann
52	CORBIS/Hulton-Deutsch Collection
53	Flip Nicklin/Minden Pictures
54	© Layne Kennedy/CORBIS
56 (left, right)	LBE
57	Photo by Michael Leahy (1933)
57 (inset)	Taken from the film *First Contact* (Australia, 1983), produced and directed by Robin Anderson and Bob Connolly

147	DD
148	LBE
149	© Kelly D. Gatlin, La Luz Photography
151, 152	LBE
153 (all images)	DD
154	Andrew Perala
155, 156	LBE
157	JPL
159	CORBIS/Archivo Iconografico, S.A.
160	CORBIS/Bettmann
161	CORBIS/Gianni Dagli Orti
162 (top)	DD
162 (bottom), 163	CORBIS/Bettmann
164 (full page, inset)	The Observatories of the Carnegie Institution of Washington
165	Courtesy of Hansen Planetarium Publications, Salt Lake City; reproduced with permission
166	AAO
168 (all photos)	LBE
169	JPL
171	French Ministry of Culture and Communication, Regional Direction for Cultural Affairs—Rhone-Alpes region—Regional Department of Archeology
172, 173	David Brill
174, 175	Erich Lessing/Art Resources, New York
177	LBE
178	Jeremy Woodhouse/PhotoDisc, Inc.
179 (left)	Tim Fitzharris/Minden Pictures
179 (right)	Flip Nicklin/Minden Pictures
180	HST, Courtesy of J. Newman (University of California at Berkeley) and NASA
181	ESO
182–183 (all photos)	Alan Sandage/Carnegie Institution of Washington
184	Copyright © 1998 by Wil Milan (http://www.astrophotographer.com)
185	Lightning photo by Michael Bath (http://australiansevereweather.simplenet.com/photography/)
188	The Bodleian Library, Oxford, MS. Marsh 139 folio 16 verso
189, 190 (top, bottom), 191	LBE
192	DD
194	John Gleason/Celestial Images
195	John Gleason/Steve Mandel
196	LBE

197	© Kelly D. Gatlin, La Luz Photography
200, 201, 202, 203, 204, 205, 206, 207	LBE
209	HST, Courtesy of R. Williams and the HDF Team (STScI) and NASA

Abbreviations

AAO	© Anglo-Australian Observatory, photography by David Malin
DD	Don Davis / Electric Image Animation Software
ESO	European Southern Observatory
GSFC	National Aeronautics and Space Administration / Goddard Space Flight Center
HST	Hubble Space Telescope; material created with support to AURA/ST Scl from NASA contract NAS5-26555 is reproduced here with permission.
LBE	From *Life Beyond Earth,* a film by Timothy Ferris, a production of KCTS Seattle in association with PBS and Devillier Donegan Enterprises.
MSSS	National Aeronautics and Space Administration / Malin Space Science Systems
NRAO	National Radio Astronomy Observatory
NRAO/AUI	National Radio Astronomy Observatory / Associated Universities, Inc.
JPL	Jet Propulsion Laboratory / California Institute of Technology / National Aeronautics and Space Administration
JSC	National Aeronautics and Space Administration / Johnson Space Center
NASA	National Aeronautics and Space Administration
NSSDC	National Aeronautics and Space Administration / National Space Science Data Center
SOHO	Solar And Heliospheric Observatory / EIT Consortium, NASA / European Space Agency
UKS	© Anglo-Australian Observatory/Royal Observatory, Edinburgh
USGS	National Aeronautics and Space Administration / United States Geological Survey

INDEX

Page numbers in *italics* refer to illustrations.

Abell 2218, *116*
Aldrin, Edwin, *58*
algae, 38, 39
aliens, *see* extraterrestrial life
Alpha Regio, Venus, *68*
Andromeda galaxy, 164
Antarctica, 88–89, *88, 89*
 Martian meteorites in, 90, *90*
"Antennae" (NGC 4038/4039), *115*
Apollo 11, *58*
Apollo 12, *58*
Apollo 15, *26, 132*
Apollo 17, *42*
Aristotle, 158, 160
Armstrong, Neil, *58*
Arsia Mons, Mars, *81*
Ascraeus Mons, Mars, *81*
astronauts, *58, 59*
astronomical observatories, 154
 see also specific observatories
Astronomie, L' (Flammarion), *160*
atmosphere, 95
Australopithecus boisei, 172

bacteria, 38, 39, *90*
 heat-loving, 96

transported to Moon, 59
Beagle, HMS, 50, *50–51*
Bell Telephone Company, 142
Beta Pictoris, 148
Big Dipper, 148
black smokers, 96, *96,* 100
Bonneville Salt Flats, Utah, *37, 38, 42*
 Cambrian life forms in, *40*
Butler, Paul, 155

Cai Guo-Qiang, 140, *141*
Cambrian explosion, 40, *40*
Capella, 148
carbon, *112*
carbon dioxide, 64
Carlyle, Thomas, 152
cave fresco, *170*
Cepheid variable star, *164*
Chimbu people, outsider contacts with, 57
Chinese ideogram, *177*
Chryse Planitia, Mars, *74*
Circuli Lactei Planetae ("Planets of the Milky Way"), *152*
Cocos Islands, *178*

Collins, Michael, *58*
colonialism, science fiction and, 58
comets, 28
 Hale Bopp, *28*
 planet formation and, *94*
Comical History of the States and Empires of the Worlds of the Moon and Sun (Cyrano de Bergerac), *130*
communication, interstellar:
 exploration vs., 125, 128, 139
 human desire for, 125, 168
 language development and, 169–77
 species longevity and, 190–93, *191, 193*
comparative planetology, 66
Confucius, 24
Conrad, Charles, Jr., *58*
consciousness, seen as uniquely human, 205
Cook, James, 44, *44, 45,* 46, 57
Cordova, France, 194, 200
Cyrano de Bergerac, *130*
Cyrano (Rostand), *130*

Dallas, Tex., *178*
Darwin, Charles, 50–56, *52,* 99
 see also evolution
Day the Earth Stood Still, The (film), 136, *137*
DNA, 54–65, 195
 extraterrestrial life forms and, 55
Doi people, outsider contacts with, 57
dolphins, *53*
Dyson, Freeman, 194, 207–8

Eagle Nebula (M16), *110*
Earth, 12, 22, 42, *42,* 59, *126,* 127, 158, 160
 cosmic "seeding" of, 34
 early atmosphere of, 95, *95*
 energy from within, 32
 fictional invasion and colonization of, 58, *58*
 Magellan's circumnavigation of, 44
 Mars's 1997 opposition to, *72*
 molten core of, 80
 nearest star to, *126*
 origination of intelligence on, 120

Earth (*continued*)
    radio broadcasts from, 146, *147,*
        148–49, *148, 149*
    Sun's distance from, 46
    Venus's surface changes and, 68
    volcanoes on, 32, *32,* 61, 68, 80, *80*
Earth, life on:
    conditions for, 93, 99
    diversity of, 13–14
    durability of, 119
    earliest fossil evidence of, 38
    onset of, 94–95
    see also life; life, origin of
Edelman, Gerald, 203
Egyptian hieroglyphics, *174*
Einstein, Albert, *162,* 163, 164, 206
Eistla Regio, Venus, *68*
emergent properties, *178*
    intelligence as, 179, 184
    weather as, 184
Enceladus, 104, *104*
energy, 14, 22, 119
    geothermal, 96
    new planet formation and, 30
    from within planets, 32
epithelial cells, *24*
Eskimo Nebula, *112*
Europa, 14
    ice on, *99,* 100, *101–102,* 103
    possibility of life on, 100–103,
        100–102
    putative global sea of, 100, 103
European Southern Observatory, *19,*
    116, 180
Evaporating Gaseous Globules
    (EGGs), *110*
evolution, 37–60, 93, 170–75,
    172–73, 202, 204
    biological sciences and, 53

Cambrian explosion and, 40
Cook's voyages and, 44, *44, 45,* 46
DNA and, 54–56
of humans, *172–73,* 175
mutation and, 50
natural selection in, 50, 55, 203
as periods of stasis and
    innovation, 39
randomness in, 50, 56, *56*
rise of mammals and, 41
science and, 53
voyage of *Beagle* and, 50, *50–51*
extraterrestrial life:
    detecting radio broadcasts from,
        149
    DNA and, 55
    Fermi's Question and, 150–51
    as intelligent, 125, 128
    probabilities of, 201
    proposed messages to, 168, 193–97
    search for, 13, 15, 19–20, 120
    SETI and, 149, 151, 190–91, 193
    species longevity and
        communicating with, 190–93,
        191, 193
    symbolic language and, 176–77
    UFO phenomenon and, *134,*
        135–36, *137*

Fermi, Enrico, 150–51
Fermi's Question, 150–51
fireworks, *132*
flaming triangle communications
    experiment, 139–40, *139*
Flammarion, Camille, *160*
flying saucers, 135
Fontenelle, Bernard de, *160*
*Fontenelle Meditates on the Plurality of
    Worlds* (Fontenelle), *160*

galaxies, 114–16
    colliding, 115, *115,* 144
    elliptical, *116*
    and expanding space, 164, *164*
    gravitational lensing and, *116*
    spiral, 180, *180, 182*
Galileo Galilei, 130, 158–60, *158,* 164
Galileo space probe, *100, 101, 102,*
    107, *107*
Ganymede, *99*
Gendler, Robert, *22*
general theory of relativity, 163
genes, 95, 207
genetic engineering, 207
geothermal energy, 96
Goldstone Deep Space
    Communications Complex, *169*
Gott, J. Richard, III, 188–89, 195, 206
Gould, Stephen Jay, 195, 205
gravitational lensing, *116*
gravity, 156, 163
Great Red Spot of Jupiter, *108*
Great Wall of China, 140, *140*
greenhouse effect, 64
greenhouse gases, 68
Grinspoon, David, 66–68, *66*
Gula Mons, Venus, *68*

habitable zone, 61, 62, 72, 73–74, 84,
    87–88, 93, 94–95, 99, 119
    see also Mars; Venus
Hale Bopp, Comet, *28*
Harbor Branch Oceanographic
    Institution, *96*
Harry K. Brown Park, Hawaii, *61*
Hawaii, volcanoes on, *61,* 80, *80*
Hesiocaeca methanicola, 96
hieroglyphics, *174*
Ho, Edward, *168*

Homo sapiens, 172, 196
    as cosmic afterthought, 178
    evolution of, *172–73,* 175
    place of, in universe, 120
    species longevity of, 188, 207
Honey, Ben, *168*
Hornet, USS, *58*
Horowitz, Paul, 194, 201
hot springs, 96
Hourglass Nebula, *112*
Hubble, Edwin, 164, *164*
Hubble Space Telescope, *20, 62, 72,*
    106, 110, 112, 115, 116, 121,
    180
hydrogen, *121,* 142

Ibn ash-Shatir, *188*
ice zone, 99–118
    Jupiter's satellites and, 99–100,
        99–102, 103
    and satellites of Neptune and
        Saturn, 104–7, *104, 106*
    young planets and, 110–14
ideogram, Chinese, *177*
infosphere, 139–66
    Earth's radio broadcasts and, 146,
        147, 148–49, *148*
    Fermi's Question and, 150–51
    radio technology and, *140,* 141,
        146–49
    radio telescope and, 142–44, *142,*
        144, 149
intelligence, 120, 170
    as accidental, 178, 205
    Darwinian evolution and, 205
    in development of universe,
        208
    as emergent property, 179, 184
    evolution of, 174

of extraterrestrials, *see* extraterrestrial life

human survival and, 188

interferometers, *155*

interstellar communication, *see* communication, interstellar

interstellar communications networks, *192*

interstellar spaceflight, 126–27, 130–34

Io, 107, *107*

Irwin, James B., *132*

Jackson Lake, Wyo., *136*

Jansky, Karl, 142, *142*

Jet Propulsion Laboratory, 74

Johnson Sea Link, *96*

Jupiter, 14, 107, 108, 156, 163
  Great Red Spot of, *108*
  satellites of, 99–101, *99–102*, 103
  speculation about life on, 108, *108*
  sunlight received by, 99

Kalapana, Hawaii, *61*

Kauffman, Stuart, 195, 204

Keck observatory, *155*

Kepler, Johannes, 130, *162, 163*

Kitt Peak National Observatory, *110*

language, 169–77, 204

Large Magellanic Cloud, *34*

lesser flamingo, *178*

life:
  as accident, 178
  in Antarctica, 88–89, *88*
  basic ingredients of, 14, 22
  cold-loving, 96
  cosmic processes and, 34
  DNA as blueprint for, 55

on Earth, *see* Earth, life on

extraterrestrial, *see* extraterrestrial life

in habitable zone, 61; *see also* habitable zone; Mars; Venus

intelligence and, 120

Martian fossils and, 90, *90*

onset of, 201

see also ice zone

life, origin of, 93–98
  conditions on Earth for, 93
  early atmosphere and, 95
  heat-loving bacteria and, 96
  potential habitats of, 108–16
  single ancestor for, 95
  thermal vents and, 96, *96*

Life Beyond Earth (film), 146

lightning, *184*

Lincoln, Abraham, 175

Lowell, Percival, *72, 73*

Lowell Observatory, Ariz., *72*

Macrosystus pyriferus, 46

Magellan, Ferdinand, 44

Magellan spacecraft, 66, *66*, 68, *68*

mammals, 41, 205

Marconi, Guglielmo, *140*, 141

Marcy, Geoffrey, 156

Mariner 10, *26*

Mars, 13, 61, 72–92, *140*, 163
  and Antarctic meteorites, 90, *90*
  "canals" of, *72, 73*
  Earth's 1997 opposition to, *72*
  flaming triangle experiment and, 139–40, *139*
  as geologically dead, 80–81
  loss of atmosphere by, 86
  Lowell's map of, *72*
  orbit of, 126, *126*

Pathfinder probes to, 84, *85, 87*

polar caps of, 72, 73, 86

speculation about life on, 72, 73–74, 84, 87–88

terraforming of, 119, *119*

Viking missions to, 74, *76, 79*

volcanoes on, 80–81, *80*

water on, 72, 73, *79*, 119

Mars Global Surveyor, *74*

Mars Orbiter Camera, 81

Martin, Lock, *136*

mathematics, symbolic language and, 176

Mauna Kea, Hawaii, 80, *80*

measurement, units of, 210

Mercury, 26, *26*

meteors:
  of Aug. 10, 1972, 136, *136*
  of Oct. 9, 1992, 136, *136*

methane, 95
  ice, 96, *96*

Milky Way galaxy, 19, 20, *34*, 120
  center of, *13*
  Earth's radio broadcasts into, 146, *147*, 148–49, *148, 149*
  humans at periphery of, 120
  number of planets in, 14, 152, *152–53*
  number of stars in, 201
  radiotelescopic mapping of, 142

Minerva Terrace, Yellowstone National Park, *93*

Mizar, 148

"Monthly Discussions of All Sorts of Books, By Several Friends," *130*

Moon, 12–13, *95*, 119
  Apollo 15 mission to, *132*
  bacteria transported to, 59

solar eclipse and, *22*

water ice on, *26*

M16 (Eagle Nebula), *110*

Munida gregaria, 46

mutation, 50

National Aeronautics and Space Administration (NASA), 74

National Radio Astronomy Observatory, *142*

natural selection, 50, 55, 203; *see also* evolution

Neptune, *28*, 104, *104*

neutron stars, 144

Newton, Isaac, *162,* 163

NGC 628, *182*

NGC 1232, *180*

NGC 2237 (Rosette Nebula), *110*

NGC 4038/4039 (Antennae), *115*

NGC 4254, *182*

NGC 4321, *182*

NGC 4414, *20*

NGC 4603, *180*

NGC 5247, *182*

NGC 5364, *182*

NGC 5457, *182*

NGC 6520, *167*

NGC 7412, *182*

nitrogen, *104*

Nixon, Richard, 58

Norma constellation, *116*

North, Edmund H., *136*

Ohkuchi Harbor, Matsusaka, Japan, *132*

Olympus Mons, Mars, 80–81, *80*

Omega Centauri, *114*

organic molecules, 14, 22, 61, 119
  in new planet formation, 30

organic molecules (*continued*)
   in solar system, 20, 26, 28
     on Titan, 106
Orion, 30, *30*
oxygen, *121*
ozone layer, 86

Pace, Norm, 202
pancake dome volcanoes, 68, *68*
Papua, New Guinea, 57, *57*
Parkinson, Sydney, 46, *46*
Pathfinder, 84, *85, 87*
Pavonis Mons, Mars, *81*
Phillips-Parker, Makana, *168*
Phoenicopterus minor, 178
Pillan Patera, Io, *107*
Pioneer Venus Explorer, *62*
Pisine, Petro, 57
planetary nebulae, *112*
planets, 12, 34, 125
   comets and formation of, *94*
   of other stars, 14, 154–56
   in elliptical orbits, *162*
   energy from within, 32
   formation of, 30, 94, *94*
   ice zones of, 110–14
   intelligent life and, 128
   number of, in Milky Way, 14, 152,
     152–53
   orbiting nearby stars, 154–56
   orbits of, 46
"Planets of the Milky Way" (*Circuli
   Lactei Planetae*), *152*
plankton, 39
Pleiades, *112*
polychaete tubeworm, *96*
polynucleotides, 55
Porsche C4S, *38*
primates, 205

"Project for Extraterrestrials" (Cai
   Guo-Qiang), 140, *141*
protostar, *94*
Ptolemy, 158
Pwyll, Europa, *101, 102*

radio, radio technology, *140,* 141
radiotelescopes, 142–44, *142, 144,*
   149
random mutation, 50, 56, *56*
RCW38, *19*
Reber, Grote, 142, *142*
reductionism, science and, *184*
relativity, general theory of, 163
Renaissance, 130
Rennie, Michael, *136*
Rocky Hill Observatory, *135*
Rosette Nebula (NGC 2237), *110*
Rostand, Edmond de, *130*

Sagittarius, *167*
Salem, Mass., *134*
Sapas Mons, Venus, *68*
Saturn, 104, *104,* 106, 146, 156
   rings of, *106*
Saturn 5 rocket, *132*
science, 14–15, 120, 196
   evolution and, 53
   as historically recent, 41
   reductionist approach to,
     184
Scott, David R., *132*
Scott, Robert, 89
Search for Extraterrestrial
   Intelligence (SETI), 149, 151,
   190–91, 193
Siberia, 139
Siding Spring telescope, *30, 34*
Sif Mons, Venus, *68*

Solar and Heliospheric Observatory
   (SOHO), *22*
solar system, *20,* 44, 90, 94, 128
   habitable zone of, 61; *see also*
     Mars; Venus
   icy moons in, 99–100, *99–102,*
     103, 104–7, *104, 106*
   orbit of, in galaxy 20
   organic molecules in, 20, 26, 28
   size of, 126
   supernovas and formation of, *34*
   water in, 26, 28, *30*
Sonora Desert, Ariz., *184*
Soviet Union, 64
space interferometry mission, *157*
space travel, 126–27, 130–34
species, 50
   chemical encoding of, 55
   longevity of, 207
spiral galaxies, 180, *180; see also*
   galaxies
stromatolites, 38, *38*
sulfuric acid, 66
sulphur, *121*
Sumerian inscription, *174*
Sumrall, Anastasia, *168*
Sun, 20, 22, 34, 61, *95,* 106, 158, 160,
   205
   corona of, *22*
   Earth's distance from, 46
   eclipse of, *22*
   formation of planets around, 94
   galactic orbit of, *20*
   and size of solar system, 126,
     126
   volume of, *22*
sunflower seeds, 55, *55*
sunspots, Galileo's observation of,
   158

Supernova 1987A, *34*
Surveyor III, *58*
Suzuki, Shunryu, 15
Sydney, Australia, *184*
symbols, manipulation of, 176

Tahiti, 44, 46
Tasmania, 58
telescopes, 157–58, 163
   of Galileo, 158, *158*
   in space, 157
terraforming, 119–22
   of Mars, 119, *119*
Tesla, Nikola, *140,* 141
Thales, 24
thermal vents, 96, *96,* 103
Thespesia populnea, 46
thunderstorms, 184, *184*
time
   depicted on highway, 37–42, *37,*
     38, 39, 40, 41, 42
   interstellar communication and,
     191, 192
Titan, 106, *106*
T-lymphocyte, *90*
Trifid Nebula, *121*
Triton, 104, *104*
tubeworm, polychaete, *96*
"Twin Peaks," Mars, *87*

"UFO sightings," 135–36
ultraviolet light, 86
United States:
   space programs of, 14
   volcanoes of, 32, 80, *80*
universe, 14
   ancients' view of, 158
   expansion of, 163–64
   as "great book," 158

human place in, 120
number of stars in, 167
Upper Wright Valley, Antarctica, *88*

Valles Marineris, Mars, *74, 76*
Venera probes, 64
    Venera 13, *65, 66, 68*
    Venera 14, *68*
Venus, 13, 61–71, 74, 80, 163
    cloud cover of, *62*
    possibility of life on, 61, 62
    radar image of, *66*
    surface of, 64, 66, 68
    transit of, 44, 46

Venera probes of, 64, *65,* 66, *68*
    volcanoes on, 68, *68*
Very Large Array (VLA)
    radiotelescope, *149, 197*
Very Large Telescope, *19, 180*
Viking space probes, 74–76, 81
Vittoria, 44
volcanoes, 32, *32,* 61
    on Earth, 32, *32,* 61, 68, 80, *80*
    on Hawaii, 80, *80*
    on Io, 107, *107*
    on Mars, 80–81, *80, 81*
    pancake, 68, *68*
    on Triton, *104*

on Venus, 68, *68*
Voyager probes, 107, *129*
    phonograph record sent on, 128, *128*
    trajectory of, *128*
    images taken by, *28, 99, 104, 106, 108, 128*

Wahkeenah Falls, Wash., *24*
War of the Worlds (Wells), 58, *58*
water, 14, 22, 24, 119
    on Mars, 72, 73, *79,* 119
    on Moon, *26*
    in new planet formation, 30

in solar system, 26, 28, *30*
    see also ice zone
weather, 184, *184*
Welles, Orson, 58, *58*
Wells, H. G., 58, *58*
White Chapel of Karnak, *174*
Wise, Robert, *136*
Worden, Alfred M., *132*

Yellowstone Caldera, 32, *32*
Yellowstone National Park, Minerva
    Terrace in, *93*

"Zen Drum" synthesizer, *66*

ABOUT THE AUTHOR

TIMOTHY FERRIS is the author of ten books on astronomy, physics, and the history and philosophy of science. He produced the Voyager phonograph record, launched by NASA aboard the twin Voyager interstellar spacecraft. In 1975 he proposed the "Interstellar Internet" hypothesis, suggesting that long-term communications among advanced civilizations would best be handled by an automated galactic network. A Guggenheim fellow and recipient of the American Institute of Physics prize, Ferris has taught five disciplines at four universities. He is currently emeritus professor at the University of California, Berkeley.